Mole Concepts and Stoichiometry:

A Chemistry Workbook for Success

Dharshi Bopegedera, Ph. D.

The Evergreen State College, Olympia, WA 98505

Published by Linus Learning

Ronkonkoma, NY 11779

ISBN 10: 1-60797-744-3

ISBN 13: 978-1-60797-744-5

Printed in the United States of America.

This book is printed on acid-free paper.

Print Number   5   4   3   2   1

TABLE OF CONTENTS

To the Student: How to Use this Workbook.. v

To the Instructor: Why Use this Workbook.. vi

WORKSHOP 1

Average Atomic Mass 1

WORKSHOP 2

The Mole 11

WORKSHOP 3

The Mole Applied to Compounds... 15

WORKSHOP 4

The Molar Mass of Elements... 19

WORKSHOP 5

The Molar Mass of Compounds 23

WORKSHOP 6

There and Back Again: Particles $\leftrightarrow$ Moles $\leftrightarrow$ Mass of Atoms....................... 27

WORKSHOP 7

There and Back Again: Particles ⟷ Moles ⟷ Mass of Molecules 35

WORKSHOP 8

Percent Composition by Mass 43

WORKSHOP 9

Balancing Chemical Equations 53

WORKSHOP 10

Reaction Stoichiometry in Moles 65

WORKSHOP 11

Reaction Stoichiometry in Grams 75

WORKSHOP 12

Limiting Reactants 85

WORKSHOP 13

Concentrations of Solutions (Molarity) 97

WORKSHOP 14

Quantitative Dilution 105

Appendix 124

To the Student: How to Use this Workbook

The workshops in this workbook were developed over the course of several years to help my first-year chemistry students learn mole concepts and stoichiometry. They were the result of over two decades of trying to teach these topics in lectures and noticing time and again that, despite my best efforts, many students struggled to grasp these important concepts. My own observations were confirmed by many chemistry professors around the world in every type of academic institution - we all recognize that the mole concepts and stoichiometry present the first stumbling block to the beginning chemistry student.

Therefore, I took a risk and tried a new approach in my class by creating **workshops that relate mole concepts and stoichiometry to students' everyday life experiences.** I also designed them so that students could learn the concepts by working through the worksheet rather than expect the professor to impart the knowledge. The results surprised me, especially when students started *asking* for more workshops! It is my hope that this workbook will serve to help *you* learn these concepts so you can be successful in your academic endeavors.

To achieve optimum results from this workbook:

1. Do these workshops in groups of two or three (optimal), being careful to form groups with those who learn at your pace. Attempt each problem individually first, then discuss your challenges. Ask questions and help each other learn, as this will enhance your own understanding. If you are moving faster or slower than your groupmates, find a different group. Having said that, some of my students have opted to do the worksheets on their own, quite successfully! Choose what works for you.

2. Do not rush through the worksheets or skip problems (or even parts of problems). It takes time to learn new things. Focus on learning the concepts instead of getting the "right" answer.

3. After completing each worksheet, reinforce your learning by doing problems from your textbook or on-line sources. It is easy to find extra problems by borrowing a chemistry book from your library written by a different author than the one you own. Your instructor may have a useful web resource for your class.

4. Return to the worksheets when you need a refresher. Reviewing reinforces learning, and it goes faster the second (or third, or....) time around.

5. You are provided room in this workbook to write down your work as you solve problems so that when you need to review, you will have easy access to the materials in one place. Even if you are working in teams, you should write down answers in your own words since that will enhance your learning.

Solutions are not included in this workbook because the goal is to encourage you/your group to figure out answers together.

Feedback Received from Students About the Workshops:

♦ *Starting small, using the dozen as an example, we learned how to work with the mole. This approach helped make the new concept less threatening and easier to understand.*

♦ *Some topics I struggled with in the past were conversions, the mole concept, and bonding. After these workshops, I wondered why I had problems with these concepts in the first place! The way the workshops were presented worked with my learning style.*

♦ *The most complex areas of general chemistry involve stoichiometry. The workshops helped me understand the concepts behind the calculations. Before taking this class, stoichiometry was mindless arithmetic, but she taught me how to do chemistry.*

♦ *Two things I know how to do really well after taking working through the worksheets - converting things to moles and balancing equations.*

To the Instructor: Why Use this Workbook

Those of us who teach first year chemistry are familiar with the challenges students face when learning mole concepts and stoichiometry. Students' lack of understating of these concepts have led to a failure rate of at least 30% in college level first-year chemistry courses,[1-3] preventing students from pursuing their academic goals. A chemist, in expressing his frustration that the mole concepts is "destroying many people's enjoyment of a splendid subject,"[4] suggested that it should be X-rated! Given the many articles published on the challenges of teaching and learning the mole and stoichiometry concepts all over the world (of which a fraction is cited here),[4-36] it is clear that chemistry faculty are eager for a better outcome, especially because mastering these concepts is key to students' further success in chemistry (about 14% of the American Chemical Society (ACS) General Chemistry Standardized Exam is on these topics). Since successful completion of first-year chemistry is required for most science and engineering majors, we have the potential to positively impact a very large number of students.

Chemistry educators have established that conceptual understanding of key topics lead to long-term success compared to drilling exercises.[37-43] The literature shows that student-centered, active learning modalities and collaborative problem-solving strategies that foster critical thinking skills are the key to success.[44-51] With that in mind, chemistry educators have developed alternative pedagogies to lectures[52-70] to facilitate student learning. Student response devices such as clickers[71-74] have also been used to improve student engagement and learning, especially in large enrollment chemistry courses. Some of these approaches require radical changes to the learning environment that may not be possible or desirable in every educational setting. Other approaches require students to invest a significant amount of time outside of class to make them effective. Challenges to these approaches have also been reported.[75]

Desiring for a better outcome for my students, I took a risk by developing my first worksheet on the average atomic mass following the principle of *teaching with your mouth shut*.[76] Since this approach necessitates that students learn concepts from the worksheet by working together in teams with minimal instructor interaction, I created the worksheet by connecting the average atomic mass to students' life experiences. While students worked in teams of three on the worksheet, I observed, answered questions (by leading to answers by asking them questions), and encouraged them to find solutions by working together. I asked a student volunteer to work each problem on the board *after* sufficient time was given for everyone to try the problem first. *Following this*, I initiated a class discussion to address challenges and misconceptions. Students were instructed to proceed to the next question only after this discussion. Homework was assigned from the textbook at the end of the workshop.

I was surprised by how well this approach worked, especially when *students asked for more worksheets*! I witnessed that with a carefully crafted worksheet, students learned and retained the information more effectively as evidenced by their exams and self-evaluations. In contrast to previous findings in chemistry education[77], this group of students performed better in ACS exams on the mole concept and stoichiometry questions (average=70%, std. dev.=25, n=31) than their average achievement on the entire exam (average=46%, std. dev.=18, n=31). I have since replaced all of my lectures on mole concepts and stoichiometry with these worksheets.

These worksheets have been:

- **used successfully with first-year college chemistry students (general chemistry and introductory chemistry)**
- **can be used in any educational setting without changes to the learning environment**
- **use collaborative problem solving strategies and active learning modes**
- **built on the premise that students learn better in the context of already acquired life experiences[78, 79]**
- **enable students to develop a conceptual understanding of the topics instead of compartmentalizing the learning of chemistry separate from other life experiences[80]**

- **using them to teach the material covered here does not take any more time than traditional lectures while being more effective in helping students learn**

Why are these worksheets effective?

1. Learning takes place in class when students have access to resources (peers, the instructor, and tutors) so misconceptions can be immediately clarified and connections made with previously learned concepts. This approach has been demonstrated to help with retention, attentiveness, and transitioning newly learned concepts from short-term to long-term memory[3, 81]

2. Students have time to take good notes since they are not rushing through the material

3. Students form friendships with teammates and connect with them outside of class for study groups and doing homework. Beyond academic help, this gives students a sense of community

4. Since most groups stay together for all the workshops, there is a sense of accountability. Students know that they will be missed if they are absent to class. They already have a group of peers to seek help in case they are absent

5. Worksheets emphasize understanding of concepts rather than algorithmic solutions

6. Worksheets provide multiple opportunities for the instructors to observe and connect with students. Therefore, the instructor can advise students more effectively, intervene if needed, and provide extra support if needed. The trust formed between the students and the instructor is carried through to other activities of the course (lectures and labs), resulting in students taking ownership of the course material instead of being passive observers[81]

Intended Audience: These workshops can be used in any chemistry course that explores the mole concepts and stoichiometry. This includes students in **high schools (introductory chemistry, honors chemistry, and AP chemistry), community colleges, liberal arts colleges, and universities (preparatory chemistry, introductory chemistry, general chemistry, and honors chemistry).** They could be recommended for more **senior students** as a self-study tool to review the concepts presented here. They are an excellent tool for training **future teachers and teaching assistants**.

The worksheets cover chemistry topics in the order they appear in most general chemistry textbooks. Following are some suggestions on how best to use the worksheets based on my experience.

1. Students should work in small teams (three is optimal) to complete the workshops. **I have found that the most effective teams are formed with students who work at a similar pace.** In my experience, students enjoy, rather than resent, learning with others who work at a similar pace and overall have a better experience. While it can be challenging to form such groups at first because students don't know each other, groups that stay together for the duration of the course emerge within a few class meetings. I encourage students to switch groups if needed (even during class) so that they can be in a team of compatible peers. In large enrollment courses taught in large lecture halls, students can pair up to complete each problem and then discuss their solutions/challenges with the pair sitting behind/in front of them. Before long, pairs that want to work as a team will sit together! The instructor can encourage this by inviting students to move around to form teams.

2. Some of my students have done these worksheet on their own, quite successfully!

3. After each problem is completed, a discussion is initiated by the instructor to clarify misconceptions and address challenges.

4. Some groups may finish their work before others. I always have extra problems, usually from their textbook, to assign to these groups. Students do not "finish and leave early" but have the opportunity to get a head start on their homework to further reinforce concepts while having access to fellow students and the instructor for needed support.

5.   I often encourage the shy students to solve problems on the board (speaking is not needed). Given this opportunity, they earn their peers' respect and are empowered by the experience.

6.   If I need to establish a definition or drive home a particularly challenging concept, I stop the workshop to address the entire class.

7.   Studies have suggested that in order to be challenged, high-achievers need complex tasks while lower-achieving students need structured help.[17] Therefore, if you have a class of well-prepared students (honors course for example), some of the workshops could be assigned as homework. In this case, class time could be used to compare work in teams, discuss challenges, and clarify misconceptions with instructor intervention.

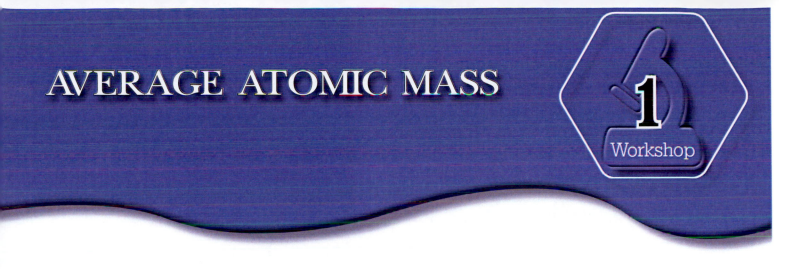

AVERAGE ATOMIC MASS

Please complete each part of each question before proceeding forward. Write down your problem solving strategy; not just the answers.

Look through your textbook and notes and write definitions for the following.

1.   Atomic number

2.   Mass number

3.   What are the symbols for atomic number and mass number?

4.   Write the <u>complete</u> symbol for the element with the atomic number 8 and the mass number 15

Now complete the following table making sure that you understand the information presented therein. The first problem is done for you.

Table 1.1

element symbol	Z	number of neutrons	A	complete element symbol
C	6	6	12	$^{12}_{6}C$
C		7		
				$^{79}_{35}Br$
	35	46		

Note that isotopes of the same element have the same value for Z but different values for A.

Different isotopes of the same element have different atomic masses. Atomic masses are measured in the laboratory using mass spectrometers (relative to the mass of the ^{12}C isotope). Atomic masses of all isotopes can be obtained from a reference book (such as the CRC Handbook of Chemistry and Physics). The unit for atomic mass is "atomic mass units" (abbreviation is: amu). See Table 1.2 for examples.

Table 1.2

isotope	atomic mass	isotope	atomic mass
$^{12}_{6}C$	12 amu (exactly)	$^{24}_{12}Mg$	23.98504 amu
$^{13}_{6}C$	13.00335 amu	$^{25}_{12}Mg$	24.98584 amu
$^{27}_{13}Al$	26.98153 amu	$^{26}_{12}Mg$	25.98259 amu

The *average atomic mass* is different from the atomic mass, although they both have the same units (amu). When you read the mass of an element from the periodic table, you are reading the *average atomic mass*. For example, look up the element carbon in the periodic table.

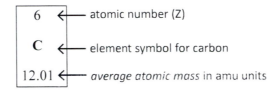

Notice that the *average atomic mass* of C is different from the atomic mass of any of its isotopes (See Table 1.2). Why is that? You will understand the reason by completing the following problems.

5. There are four students in a group. Their ages are 21, 19, 28, and 24 years respectively.

   a) What is the average age of the student group?

   b) Do any of the students have this "average" age?

   c) What is the purpose of calculating the average age?

6. The age of 20 students in a class are given in the following table.

number of students	Age (yrs)
2	32
6	20
3	21
1	50
5	22
3	19

a) Calculate the average age.

b) Do any of the students have this "average" age? Why?

c) What is the purpose of calculating the average age?

7. The ages of students in a class are given in the following table.

students	Percentage	Average age (yrs)
male	36.21%	24
female	63.79%	20

a) Estimate (without calculating) the average age of the students in the class. Give reasons for your answer.

b) Calculate the average age. Compare it with your estimate.

c) What is the purpose of calculating the "average"? Why is the "average" useful?

d) Why is this data represented as a percentage?

e) How many students are actually in the class? Does it impact your answer to part b above? Why or why not?

f) The average age you calculated for this class is called a *weighted average* because it is dependent on (or weighted by) the percentage of males and females in the class (weighted by population). Is the average age weighted more towards the males or the females? Why?

8. The element Cu has two isotopes ^{63}Cu and ^{65}Cu. Their natural abundances (how they are found in nature) and masses are given below. What is the average atomic mass of copper? Compare your answer with that in your periodic table.

isotope	abundance	atomic mass
^{63}Cu	69.09%	62.93 amu
^{65}Cu	30.91%	64.93 amu

Note that this problem is not very different from Problem 7 above. In Problem 7, the population of each gender is given as a percentage. Here the abundance of each isotope of Cu is given as a percentage.

Would you describe the average atomic mass you calculated for copper as a weighted average? Why?

9. The element B has two isotopes ^{10}B and ^{11}B. Their natural abundances and masses are given below. What is the average atomic mass of B? Is this a weighted average? Why or why not? Compare your answer with that in your periodic table.

isotope	abundance	atomic mass (amu)
^{10}B	19.9 %	10.012937
^{11}B	80.1 %	11.009305

10. The element Mg has three isotopes. Their natural abundances and masses are given below. What is the average atomic mass of Mg? Compare your answer with that in your periodic table.

isotope	abundance	atomic mass (amu)
^{24}Mg	78.99 %	23.985042
^{25}Mg	10.00 %	24.985837
^{26}Mg	11.01%	25.982593

11. The element Ge has five isotopes. Their natural abundances and masses are given below. What is the average atomic mass of Ge? Compare your answer with that in your periodic table.

isotope	abundance	atomic mass (amu)
^{70}Ge	20.84 %	69.924250
^{72}Ge	27.54 %	71.922076
^{73}Ge	7.73 %	72.923459
^{74}Ge	36.28%	73.921178
^{76}Ge	7.61%	75.921403

12. The element Ni has the following isotopes and natural abundances. What is the average atomic mass of Ni? Compare your answer with that in your periodic table.

isotope	abundance	atomic mass (amu)
^{58}Ni	68.0769 %	57.935348
^{60}Ni	26.2231%	59.930791
^{61}Ni	1.1399%	60.931060
^{62}Ni	3.6345 %	61.928349
^{64}Ni	0.9256%	63.927970

13. Cobalt has only one naturally occurring isotope ^{59}Co. It has an atomic mass of 58.933200 amu. What do you expect the average atomic mass to be, and why?

14. Describe the difference between the atomic mass of the ^{63}Cu isotope and the average atomic mass of copper in your own words.

Summary:

♦ The average atomic mass it is a weighted average, weighted by percent abundance of the isotopes.

♦ The periodic table provides the average atomic mass for each of the elements.

♦ If you need to know the atomic mass of a single isotope of a given element, use a reference book (such as the CRC Handbook of Chemistry and Physics) to look it up.

The problems above are helpful in learning the concept of the average atomic mass. Be sure to practice more problems from your chemistry textbook to master this concept.

Please complete each part of each question before proceeding forward. Seek help if/when you need. **Write down your problem solving strategy; not just the answers.**

You have 360 marbles. You are asked to determine how many dozens of marbles you have. Set up the problem as follows.

$$360\,marbles\,x\left(\frac{1\,dozen\,marbles}{12\,marbles}\right)=30\,dozen\,marbles$$

Notice that by setting up the problem as above, you can cancel "marbles" from the numerator and denominator (as shown below) and your final units are "dozen marbles." Note that units are just as important as the numerical answer.

$$360\,\cancel{marbles}\,x\left(\frac{1\,dozen\,marbles}{12\,\cancel{marbles}}\right)=30\,dozen\,marbles$$

Strategy: Note that you are starting the problem with what you know (i. e. 360 marbles) on the left hand side of the above equation and proceeding to what you need (i. e. how many dozens of marbles?) on the right hand side. The conversion factor (1 dozen marbles/12 marbles) is used to convert what you have (i.e. 360 marbles) to obtain what you need (i. e. how many dozens of marbles?)

Now set up the following problems similarly and solve. The first problem is done for you.

1. How many dozens of Fe atoms are there in 240 Fe atoms?

$$240\,\cancel{Fe\,atoms}\,x\left(\frac{1\,dozen\,Fe\,atoms}{12\,\cancel{Fe\,atoms}}\right)=20\,dozen\,Fe\,atoms$$

2. How many dozens of Cu atoms are there in 9875 Cu atoms?

3. You have 312.6 dozen titanium atoms. How many atoms do you have?

Chemists do not count particles using the dozen (bakers do, and so do we all in everyday life). Chemists count particles mostly in **moles** because they are interested in counting tiny particles (such as atoms, molecules, electrons, etc.). Counting by dozens does not make sense because we will have to assemble many trillions of dozens of particles before we can actually see that sample. So we use the mole instead. Mole is our **unit of choice** when counting microscopic particles such as atoms and molecules.

Mole is a counter (similar to the dozen). Let's see if we can us the same principle as above to work with the concept of the mole.

one dozen = 12 (a small number)

one mole = 6.02x10²³ (a very large number) [mol is the abbreviation for mole]

one mole = 6.02x10²³ = Avogadro Number (in honor of Amedeo Avogadro)

So we can say:

- One dozen pencils = 12 pencils
- One dozen C atoms =12 C atoms
- One dozen H_2O molecules = 12 H_2O molecules
- One mole pencils = **6.02x10²³ pencils**
- One mole C atoms = **6.02x10²³ C atoms**
- One mole H_2O molecules = **6.02x10²³ H_2O molecules**

Notice that regardless of the type of item (pencils, C atoms, H_2O molecules) as long as we are referring to a dozen, there are 12 of those items. Similarly, regardless of the type of item (pencils, C atoms, H_2O molecules) as long as we are referring to a mole, there are 6.02x10²³ of those items.

4. How many Fe atoms are there in 2 mol of Fe?

$$2\,mol\,Fe\,atoms\,x\left(\frac{6.02x10^{23}\,Fe\,atoms}{1\,mol\,Fe\,atoms}\right)=1.204x10^{24}\,Fe\,atoms$$

Strategy: Start from what you know (on the left hand side). Then use a conversion factor to obtain what you need as you proceed to the right hand side.

5. How many W atoms are there in 12.232 mol of W?

6. How many H_2O molecules are there in 2 mol of H_2O?

7.   How many CO_2 molecules are there in 12.232 mol of CO_2?

8.   How many moles are there in 2.3 billion NH_3 molecules?

9.   If you have 0.456 moles of H atoms, how many atoms do you have?

10.   How many moles of boron are present in a sample of 3.459×10^{25} boron atoms?

11.   How many atoms of Al are there in 2.23 moles of Al?

12. If you have 3.48x10^{-5} moles of pencils, how many pencils do you have?

13. Which has more atoms and why?

  a)   1 mole of Se or 1 mol of Sb?

  b)   3 mol of Au or 2.5 mol of Be?

  c)   2.68 mol of Pb or 2.68 mol of Ag?

The mole (or Avogadro number) is a very important concept in chemistry. Chemistry students are expected to know the value of the Avogadro number (6.02x10^{23}).

Be sure to revisit this Workshop when you need to review this topic. Please practice more problems from your chemistry textbook to master this concept since it is a useful concept that will be used in the rest of this workbook.

Please complete each part of each question before proceeding forward. Write down your problem solving strategy; not just the answers.

The first few problems are done for you.

Remember that **one mole = 6.02x10²³ = Avogadro number**

1. Your hand is made of 4 fingers and one thumb. How many fingers and thumbs are there in:

    a) 1 dozen hands?

    $$1 \text{ dozen hands} \times \left(\frac{12 \, hands}{1 \, dozen \, hands} \right) = 12 \text{ hands}$$

    $$12 \text{ hands} \times \left(\frac{1 \, thumb}{1 \, hand} \right) = \underline{\underline{12 \text{ thumbs}}}$$

    You can also set this up in one line as follows.

    $$1 \text{ dozen hands} \times \left(\frac{12 \, hands}{1 \, dozen \, hands} \right) \times \left(\frac{1 \, thumb}{1 \, hand} \right) = \underline{\underline{12 \text{ thumbs}}}$$

    Similarly:

    $$1 \text{ dozen hands} \times \times \left(\frac{12 \, hands}{1 \, dozen \, hands} \right) \times \left(\frac{4 \, fingers}{1 \, hand} \right) = \underline{\underline{48 \text{ fingers}}}$$

    Strategy: Start from what you know (on the left hand side). Then use one or more conversion factors to obtain what you need as you proceed to the right hand side.

    b) 1 mol of hands?

    $$1 \text{ mole hands} \times \left(\frac{6.02 \times 10^{23} \, hands}{1 \, mole \, hands} \right) \times \left(\frac{1 \, thumb}{1 \, hand} \right) = \underline{\underline{6.02 \text{x} 10^{23} \text{ thumbs}}}$$

    $$1 \text{ mole hands} \times \left(\frac{6.02 \times 10^{23} \, hands}{1 \, mole \, hands} \right) \times \left(\frac{4 \, fingers}{1 \, hand} \right) = \underline{\underline{2.45 \text{x} 10^{24} \text{ fingers}}}$$

    c) 2.5 moles of hands?

d)   0.36 mol of hands?

2.   In one molecule of CH_4:

    a)   How many C atoms are there?

    b)   How many H atoms are there?

Note that the CH_4 molecule (with one carbon atom and four hydrogen atoms) can be easily compared with your hand (one thumb and four fingers). If you have challenges with the following problems, go back to the previous problem or use your hand for comparison.

3.   In one mole of CH_4:

    a)   How many C atoms are there?

    b)   How many H atoms are there?

4.   In 2.5 mole of CH_4:

    a)   How many C atoms are there?

b)   How many H atoms are there?

5.   In 3.2 mole of C_2H_5OH:
a)   How many H atoms are there?

b)   How many O atoms are there?

6.   In 0.346 mole of $C_6H_{12}O_6$:
a)   How many H atoms are there?

b)   How many O atoms are there?

c)   How many C atoms are there?

7.   Calcium carbonate ($CaCO_3$) is the active ingredient in some most antacid tablets. How many oxygen atoms are there in 1.67 moles of calcium carbonate?

8. How many oxygen atoms are there in 2.12 moles of calcium phosphate $[Ca_3(PO_4)_2]$?

9. CFC-12, with the molecular formula CCl_2F_2, is a common refrigerant. Determine the number of chlorine atoms in 3.14 moles of this chemical.

10. Aluminum metal is extracted from the mineral cryolite (Na_3AlF_6). Determine the number of fluorine atoms in 6.32 moles of cryolite.

Make sure you practice more problems from your chemistry textbook to master this concept since it is essential for your understanding of future topics in this workbook.

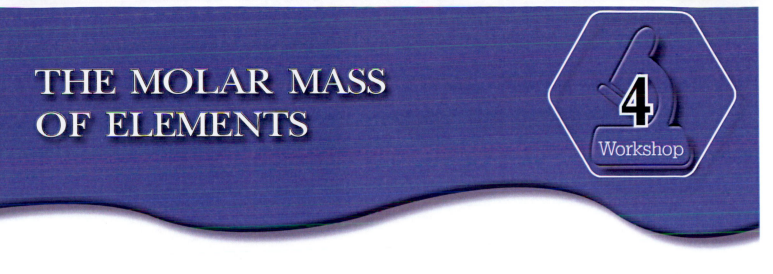

THE MOLAR MASS OF ELEMENTS

Please complete each part of each question before proceeding forward. Seek help if/when you need. **Write down your problem solving strategy; not just the answers.**

The first few problems are done for you.

1. The mass of one carbon atom is 12.01 amu. What is the mass of 10 C atoms?

$$10\,\cancel{C\,atoms}\ x\frac{12.01\,amu}{1\,\cancel{C\,atom}}=\ 120.1\,amu$$

2. What is the mass of 3.5 dozen C atoms?

$$3.5\,\cancel{dozen\,C\,atoms}\ x\frac{12\,\cancel{C\,atoms}}{1\,\cancel{dozen\,C\,atoms}}\ x\frac{12.01\,amu}{1\,\cancel{C\,atom}}=504.42\,amu$$

   Strategy: In the above two problems, as in previous workshops, start from what you know (on the left hand side). Then use one or more conversion factors to obtain what you need as you proceed to the right hand side.

3. What is the mass of:

   a) 1 mol of C atoms?

   b) 2.34 mol of C atoms

c) Note the units used in the above calculations. Will these units work in the laboratory? What units of mass are you likely to use in the laboratory?

Now let's calculate the mass in units that we are likely to use the laboratory – grams.

4. The mass of one copper atom is 63.55 amu. Determine the mass of one mole of Cu atoms first in amu and then in grams (1 amu = $1.66053878 \times 10^{-24}$ g).

5. Similarly, determine the mass of one mole of each of the following elements first in amu and then in grams.

a) Ag

b) Fe

c) N

6. Now complete the following table.

element	mass of 1 atom (amu/atom)	mass of 1 mol of atoms (g/mol)
Cu		
Ag		
Fe		
N		

Summarize your results:

If the mass of <u>one atom</u> of an element = x amu/atom, then the mass of <u>one mole of atoms</u> of

that element = _____ g/mol

Mass of one mole is called **molar mass** or **molecular weight** and the units are **grams/mol**.

Complete the following table by using **only** your periodic table.

7. Mass of a P atom =	8. Molar mass of P =
9. Mass of a K atom =	10. Molar mass of K =
11. Molar mass of Mg =	12. Molar mass of Sr =

I hope you experienced that you can use the periodic table to find the atomic mass of any element (units are amu/atom) or the molar mass of any element (units are grams/mol). This is yet another useful piece of information you can get from the periodic table.

13. Read the following definition of the mole by the International Union of Pure and Applied Chemists (IUPAC).

The mole is the amount of substance of a system which contains as many elementary entities as there are atoms in 0.012 kilogram of carbon 12; its symbol is "mol." When the mole is used, the elementary entities must be specified and may be atoms, molecules, ions, electrons, other particles, or specified groups of such particles. ("The International System of Units (SI)" Editors: Barry N. Taylor, Ambler Thompson, National Institute of Standards and Technology Special Publication 330, 2008 Edition, pp 21)

Discuss in your group the connection between this IUPAC definition and what you learned in your mole workshops. Write down your understanding in your own words.

Make sure you practice more problems from your chemistry textbook to master this concept since it is essential for your understanding of future topics in this workbook.

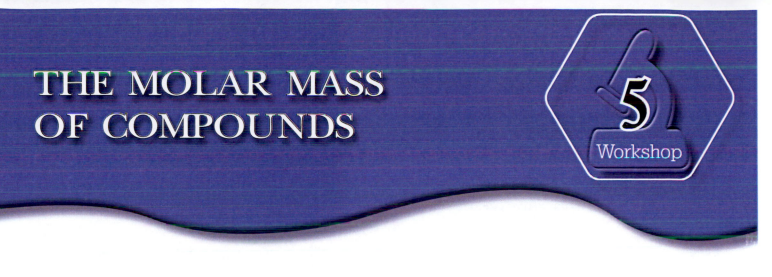

Please complete each part of each question before proceeding forward. Seek help if/when you need. **Write down your problem solving strategy; not just the answers.**

Now that you understand molar mass of elements, you are ready to study the next topic, the molar mass of compounds. If you need, please review the previous workshop.

1. CH_4 molecule has one C atom and four H atoms.

    a. Mass of one CH_4 molecule = mass of 1 C atom + mass of 4 H atoms =

    _____ + _____ = _____ [units?]

    b. Mass of one dozen CH_4 molecule = mass of _____ C atoms + mass of

    _____H atoms = _____ [units?]

    c. Mass of one mol CH_4 molecule = mass of _____ C atoms + mass of _____H atoms =
    _____ + _____ [use the same units as above for comparison]

The above answer should be in amu/mol units. Now convert it to g/mol units (1 amu = 1.66053878 x 10^{-24} g)

Note that the mass of one CH_4 molecule is 16.042 amu (or amu/molecule). The mass of one mole of CH_4 is 16.042 g/mol. As we noted in Workshop 4, the mass of one mole is called the molar mass. Therefore, the **molar mass of CH_4 is 16.042 g/mol.**

Combining what you learned in Workshop 4 with what you just learned in this workshop:

- **Mass of one C atom is 12.01 amu; mass of one mole of C atoms is 12.01 g/mol.**
- **Mass of one CH_4 molecule is 16.042 amu; the mass of one mole of CH_4 is 16.042 g/mol.**

As we observed in Workshop 3, the CH_4 molecule (with one carbon atom and four hydrogen atoms) can be compared with your hand (one thumb and four fingers). Think about how you would measure the mass of your hand – you will need to add up the masses of one thumb and four fingers.

- The mass of two hands = mass of 2 x (one thumb + four fingers)
- The mass of six hands = mass of 6 x (one thumb + four fingers)
- The mass of a dozen hands = mass of 12 x (one thumb + four fingers)
- The mass of one mole of hands = mass of 1mole x (one thumb + four fingers)

Similarly:

- The mass of two CH_4 molecules = mass of 2 x (C atom + four H atoms)
- The mass of six CH_4 molecule = mass of 6 x (C atom + four H atoms)
- The mass of a dozen CH_4 molecules = mass of 12 x (C atom + four H atoms)
- The mass of one mole of CH_4 molecules = mass of 1mole x (C atom + four H atoms)

Let's calculate the molar mass of NH_3.

One strategy is to calculate the mass of one molecule in amu units. Then we know that the molar mass (mass of one mole) is the same value but in g/mole units.

mass of one molecule of NH_3 = mass of 1 N atom + mass of 3 H atoms = 14.01 amu + 3 x 1.008 amu = 17.034 amu or 17.034 amu/molecule
Therefore, the molar mass of NH_3 = 17.034 g/mol

The other strategy is to directly calculate the molar mass of NH_3 by using molar masses of each of the elements that make up NH_3.

molar mass of NH_3 = mass of one mole of NH_3 = mass of 1mole of (N atoms + 3 H atoms) = mass of 1 mole of N atoms + mass of 3 moles of H atoms = 14.01 g/mol + 3 x 1.008 g/mol = 17.034 g/mol

Both strategies work well although the second strategy is easier to use with some practice. Use these strategies to solve the following problems.

2.   Complete the following table.

molecule	mass of 1 molecule (amu/molecule)	mass of 1 mole of molecules (g/mol) or **molar mass** (g/mol)
SF_6		
$CaCO_3$		
H_2O		
$C_6H_{12}O_6$		
Na_2SO_4		
$Fe(OH)_2$		
$Ca(NO_3)_2$		
CH_3COOH		
$Pb(NO_3)_2$		
XeF_4		

Make sure you practice more problems from your chemistry textbook to master this concept since it is essential for your understanding of future topics in this workbook.

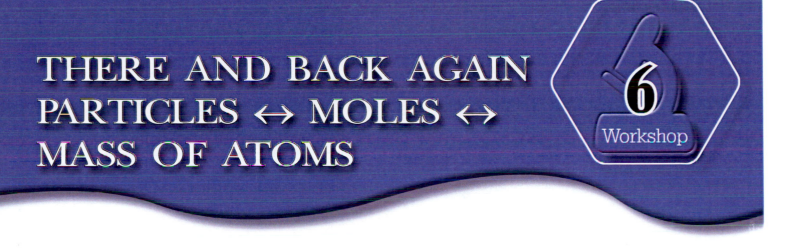

THERE AND BACK AGAIN PARTICLES ↔ MOLES ↔ MASS OF ATOMS

Workshop 6

Please complete each part of each question before proceeding forward. Seek help if/when you need. **Write down your problem solving strategy; not just the answers.**

Chemists work in the **macroscopic** world of the laboratory but want to understand the **microscopic** world of atoms and molecules. **The mole (Avogadro number) and the molar mass enables us to connect the macroscopic world of the laboratory with the microscopic world of atoms and molecules** as shown below.

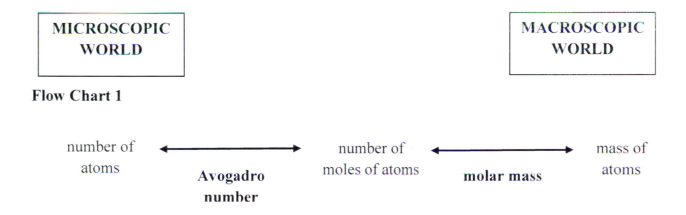

| MICROSCOPIC WORLD | | | MACROSCOPIC WORLD |

Flow Chart 1

This means that the Avogadro number is used as a conversion factor to convert between the number of atoms and moles of atoms. The <u>molar mass of elements</u> is used to convert between moles of atoms and mass of atoms.

If you are working with a group of fellow students, discuss the above flow chart with your team. Discuss how the conversion factors shown in this chart are needed to convert between the macroscopic and microscopic worlds. Make sure you understand why the chemistry laboratory is a macroscopic world and why it is important to understand the microscopic world. If you are working alone, go through this chart carefully and make sure you understand it.

Let's solve the following problem using **Flow Chart 1**.

Example: You have 6.35×10^{32} Br atoms. How many moles of Br atoms do you have?

Note: We are starting with Br atoms. We cannot see them. We are in the microscopic world!

Strategy: Set up the problem using the first half of Flow Chart 1. As in previous workshops, always start from what you know (on the left hand side) and use conversion factors to obtain what you need as you proceed to the right hand side.

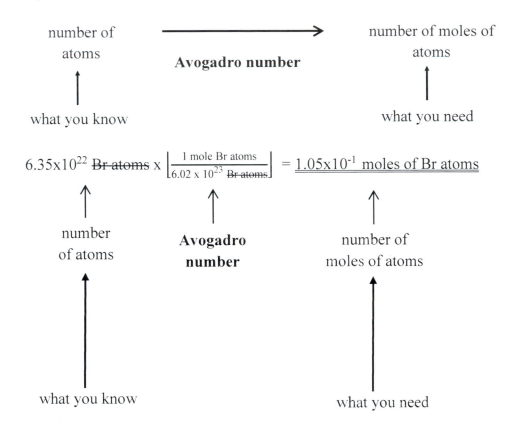

Note that the Avogadro number is the conversion factor that enables the conversion of number of atoms to moles.

Notice that in the laboratory we do not have a tool to count/measure moles. We need to find a way to relate the number of moles to something we can measure in the laboratory (the macroscopic world we work in).

To do so, calculate the mass of Br.

Strategy: Set up the problem using the second half of Flow Chart 1. As in previous workshops, always start from what you know (on the left hand side) and use conversion factors to obtain what you need as you proceed to the right hand side.

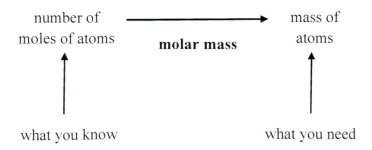

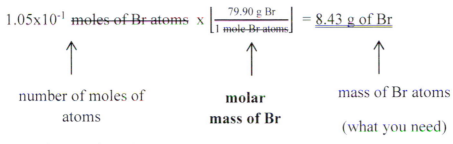

$$1.05 \text{x} 10^{-1} \text{ moles of Br atoms} \times \left[\frac{79.90 \text{ g Br}}{1 \text{ mole Br atoms}} \right] = \underline{8.43 \text{ g of Br}}$$

number of moles of atoms

(what you know)

molar mass of Br

mass of Br atoms

(what you need)

Molar mass is the conversion factor that enables the conversion of the number of moles to mass.

NOTE: We can weigh out 8.43 g of Br in the laboratory. We are in the macroscopic world! We were given data in the microscopic world ($6.35 \text{x} 10^{32}$ Br atoms) and we used the conversion factors in Flow Chart 1 to find the equivalent mass of Br atoms so we can weigh out this sample in grams in the laboratory.

Let's do a few more examples.

Example: A particular reaction requires $3.25 \text{x} 10^{21}$ sulfur atoms. Determine the mass of sulfur you will need to weigh out in the laboratory for this experiment.

You are given microscopic world information (number of atoms) and asked to determine macroscopic world information (mass). Use Flow Chart 1.

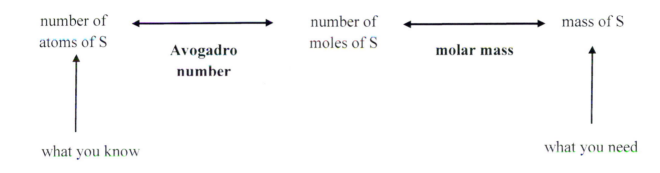

number of atoms of S

Avogadro number

number of moles of S

molar mass

mass of S

what you know

what you need

First half of Flow Chart 1:

$$3.25 \text{x} 10^{21} \text{ S atoms} \times \frac{1 mole\ S}{6.02 x 10^{23}\ S\ atoms} = \underline{5.39 \text{x} 10^{-3} \text{ mole S}}$$

Second half of Flow Chart 1:

$$5.39 \text{x} 10^{-3} \text{ mole S} \times \frac{32.07 g\ S}{1\ mole\ S} = \underline{0.173\ g\ S}$$

With a little practice, you will be able to do this all in one step as follows. Try it!

$$3.25 \text{x} 10^{21} \text{ S atoms} \times \frac{1\ mole\ S}{6.02 x 10^{23}\ S\ atoms} \times \frac{32.07 g\ S}{1\ mole\ S} = \underline{0.173\ g\ S}$$

You do not *have* to do this problem in one step. It is perfectly fine to use one step at a time to achieve your goal.

Example: In the laboratory you weigh out 22.38 g of iron. You need to know how many iron atoms are in this sample. Calculate this value.

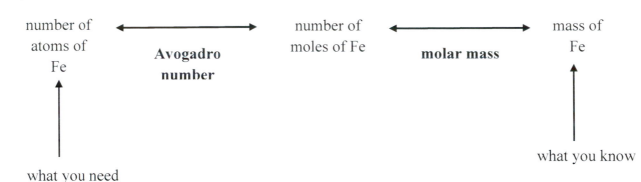

Notice that we are starting from the right hand end of this flow chart (we are given the mass of Fe) and need to work towards the left to find the desired answer (number of atoms of Fe). In other words, we are starting from the macroscopic world information and need to calculate microscopic world information.

Flow Chart 1 can be used to go from left to right or from right to left. That is why this workshop is titled "there and back again."

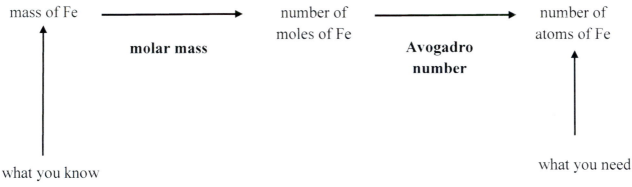

First half of the calculation:

$$22.38 \; \cancel{g \, Fe} \times = \frac{1 \, mole \, Fe}{55.85 \; \cancel{g \, Fe}} = \underline{\underline{0.4007 \; mole \; Fe}}$$

Second half of the calculation:

$$\underline{\underline{0.4007}} \; \cancel{mole \, Fe} \times \frac{6.02 x 10^{23} \, Fe \, atoms}{1 \; \cancel{mole \, Fe}} = \underline{\underline{2.412 x 10^{23} \; Fe \; atoms}}$$

You can also do this problem in one step (as shown before). Try it!

$$22.38 \; \cancel{g \, Fe} \times \frac{1 \; \cancel{mole \, Fe}}{55.85 \; \cancel{g \, Fe}} \times \frac{6.02 x 10^{23} \, Fe \, atoms}{1 \; \cancel{mole \, Fe}} = \underline{\underline{2.412 x 10^{23} \; Fe \; atoms}}$$

Please review the above examples carefully before proceeding to the following problems. Note that you may not always have to use both steps of Flow Chart 1. Depending on the problem, you may only need one half of the chart.

1. Convert each of the following into number of moles.

    a. 2.45×10^{32} atoms of N

    b. 6.26×10^{22} atoms of P

    c. 1.23×10^{21} atoms of O

    d. 8.23×10^{24} atoms of Cu

2. Convert each of the following into grams.

    a.   0.23 moles of Zn

    b.   2.98 moles of As

    c.   4.46 moles of Hg

    d.   6.94 moles of Ag

e.   5.68 moles of Pt

3.   Pure magnesium solid ignites readily when heated producing a bright white flash. If you have 3.2×10^{22} magnesium atoms, what is the mass of your magnesium sample?

4.   Diamond is a pure form of carbon. What is the weight of 2.35×10^{24} diamond atoms?

5.   You have weighed out 5.92 g of manganese metal in the laboratory. How many manganese atoms are in your sample?

6. You have been given a silicon chip that weighs 23.89 g for an experiment. You need to know the number of atoms of silicon present in this sample. Calculate this value.

7. The thermite reaction, an explosive reaction between aluminum and iron(III) oxide, requires 23.48 g of aluminum powder. Determine the number of aluminum atoms in this sample.

8. You are provided with a gaseous argon sample that weighs 43.25 g. how many argon atoms are present in this sample?

Make sure you practice more problems from your chemistry textbook to master this concept since it is essential for your understanding of future topics in this workbook.

THERE AND BACK AGAIN PARTICLES ↔ MOLES ↔ MASS OF MOLECULES

Please complete each part of each question before proceeding forward. Seek help if/when you need. **Write down your problem solving strategy; not just the answers.**

You learned in Workshop 6 that chemists work in the **macroscopic** world of the laboratory but want to understand the **microscopic** world of atoms and molecules. **The mole (Avogadro number) and the molar mass enables us to connect the macroscopic world of the laboratory with the microscopic world of atoms and molecules.** We used Flow Chart 1 to convert between the microscopic and macroscopic worlds for elements.

This workshop involves conversion between the microscopic and macroscopic world for molecules. See Flow Chart 2.

MICROSCOPIC WORLD	MACROSCOPIC WORLD

Flow Chart 2

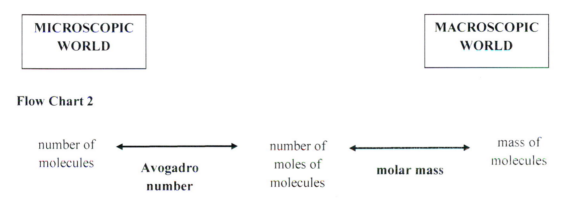

This means that the Avogadro number is used as a conversion factor to convert between the number of molecules and moles of molecules. The <u>molar mass of molecules</u> is used to convert between moles of molecules and mass of molecules.

If you are working with a group of fellow students, discuss the above flow chart with your team. Discuss how the conversion factors shown in this chart are needed to convert between the macroscopic and microscopic worlds. Make sure you understand why the chemistry laboratory is a macroscopic world and why it is important to understand the microscopic world. If you are working alone, go through this chart carefully and make sure you understand it.

Let's solve the following problem using **Flow Chart 2**.

Example: You have 9.98×10^{22} water molecules. How many moles of water do you have?

Note: We are starting with water molecules. We cannot see them. We are in the microscopic world!

Strategy: Set up the problem using the first half of Flow Chart 2. As in previous workshops, always start from what you know (on the left hand side) and use conversion factors to obtain what you need as you proceed to the right hand side.

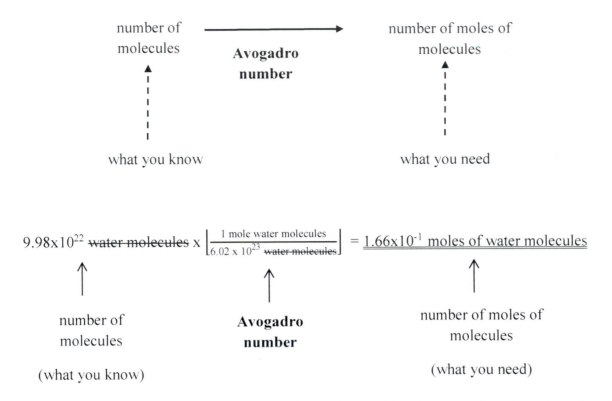

Note that the Avogadro number is the conversion factor that enables the conversion of number of molecules to moles.

Notice that in the laboratory we do not have a tool to count/measure moles. We need to find a way to relate the number of moles to something we can measure in the laboratory (the macroscopic world we work in).

To do so, calculate the mass of water.

Strategy: Set up the problem using the second half of Flow Chart 2. As in previous workshops, always start from what you know (on the left hand side) and use conversion factors to obtain what you need as you proceed to the right hand side.

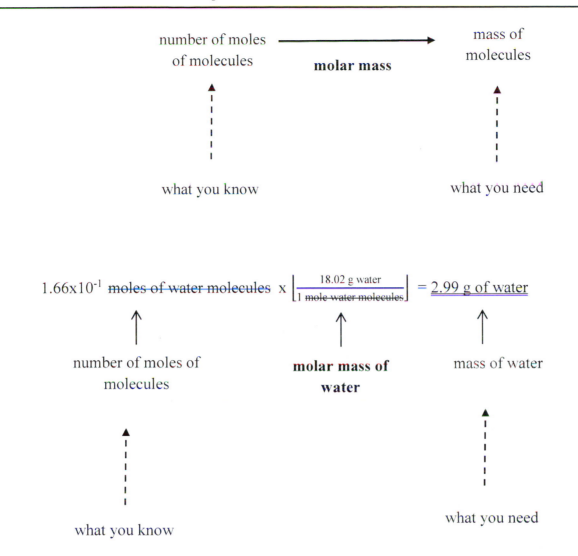

$$1.66 \times 10^{-1} \; \text{moles of water molecules} \; \times \left[\frac{18.02 \text{ g water}}{1 \text{ mole water molecules}} \right] = \underline{2.99 \text{ g of water}}$$

number of moles of molecules

molar mass of water

mass of water

what you know

what you need

Note that the molar mass is the conversion factor that enables the conversion of the number of moles to mass.

NOTE: We can weigh out 2.99 g of water in the laboratory. We are in the macroscopic world! We were given data in the microscopic world (9.98×10^{22} water molecules) and we used the conversion factors in Flow Chart 2 to find the equivalent mass of water molecules so we can weigh out this sample in grams in the laboratory.

Let's do a few more examples.

Example: A particular experiment requires 9.82×10^{28} molecules of ethanol (C_2H_5OH). Determine the mass of ethanol you will need to weigh out in the laboratory.

You are given microscopic world information (number of molecules) and asked to determine macroscopic world information (mass). Use Flow Chart 2.

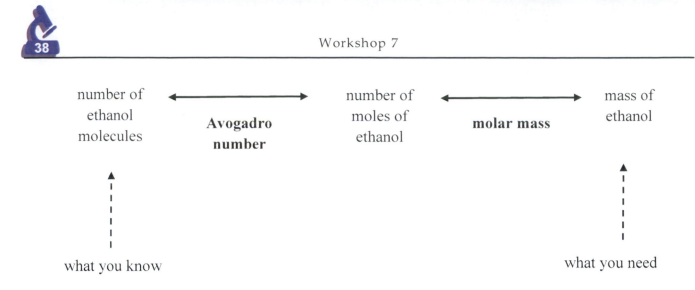

First half of Flow Chart 2:

$$9.82 \times 10^{25} \text{ ethanol molecules } \times \frac{1\,mole\,ethanol}{6.02 \times 10^{23} \text{ ethanol molecules}} = 1.63 \times 10^{2} \text{ mole ethanol}$$

Second half of Flow Chart 2:

This requires the molar mass of ethanol which is 46.068 g/mol (check it!).

$$1.63 \times 10^{2} \text{ mole ethanol } \times \frac{46.068\,g\,ethanol}{1\,mole\,ethanol} = 7.51 \times 10^{3} \text{ g ethanol}$$

With a little practice, you will be able to do this all in one step as follows. Try it!

$$9.82 \times 10^{25} \text{ ethanol molecules } \times \frac{1\,mole\,ethanol}{6.02 \times 10^{23} \text{ ethanol molecules}} \times \frac{46.068\,g\,ethanol}{1\,mole\,ethanol} = 7.51 \times 10^{3} \text{ g ethanol}$$

You do not *have* to do this problem in one step. It is perfectly fine to use one step at a time to achieve your goal.

Example: You weigh out 6.59 g of calcium carbonate ($CaCO_3$) in the laboratory for an experiment. Calculate the number of molecules of calcium carbonate in this sample.

Use Flow Chart 2.

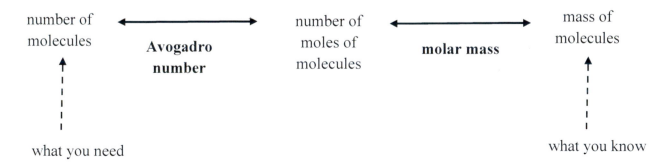

Notice that we are starting from the right hand end of this flow chart (we know the mass of $CaCO_3$) and need to work towards the left to find the desired answer (number of molecules of $CaCO_3$). In other words, we are starting from the macroscopic world information and need to calculate microscopic world information.

Flow Chart 2 can be used to go from left to right or from right to left. That is why this workshop is titled "there and back again."

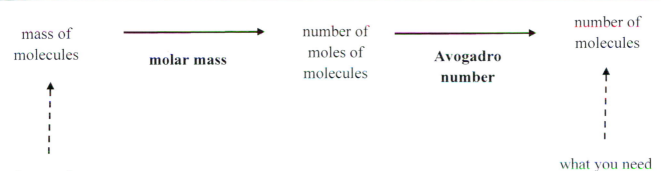

what you know

First half of the calculation: This requires the molar mass of $CaCO_3$ (100.087 g/mol). Check it!

$$6.59 \text{ g } CaCO_3 \times \frac{1 \, mole \, CaCO_3}{100.087 \, g \, CaCO_3} = 6.58 \times 10^{-2} \, mole \, CaCO_3$$

Second half of the calculation:

$$6.58 \times 10^{-2} \, mole \, CaCO_3 \times \frac{6.02 \times 10^{23} \, CaCO_3 \, molecules}{1 \, mole \, CaCO_3} = 3.964 \times 10^{22} \, CaCO_3 \, molecules$$

You can also do this problem in one step (as shown before). Try it!

$$6.59 \text{ g } CaCO_3 \times \frac{1 \, mole \, CaCO_3}{100.087 \, g \, CaCO_3} \times \frac{6.02 \times 10^{23} \, CaCO_3 \, molecules}{1 \, mole \, CaCO_3} = 3.964 \times 10^{22} \, CaCO_3 \, molecules$$

Please review the above examples carefully before proceeding to the following problems. Note that you may not always have to use both steps of Flow Chart 2. Depending on the problem, you may only need one half of the chart.

1. Convert each of the following into number of moles.

    a. 5.32×10^{23} molecules of ammonia (NH_3)

    b. 8.94×10^{22} molecules of methane (CH_4)

c. 3.53×10^{24} molecules of HCl

d. 3.68×10^{21} molecules of SF_6

e. 2.55×10^{18} molecules of XeF_4

2.   Convert each of the following into grams.

a.   1.45 moles of PCl_3

b.   12.98 moles of NH_3

c.   6.54 moles of $CHCl_3$

d.   2.18 moles of AgCl

e.   8.27 moles of $CaCl_2$

3.   Benzene (C_6H_6) is a common organic solvent. What is the mass of 2.35×10^{21} molecules of benzene?

4.   Octane (C_8H_{18}) is one of the main ingredients of gasoline. What is the mass of 8.01×10^{23} octane molecules?

5. Single-use cold packs are based on the reaction between NH_4NO_3 and water (endothermic reaction producing a cooling effect). Determine the number of NH_4NO_3 molecules in 6.26g of this material.

6. The reaction between $CaCl_2$ and water is exothermic (generates heat). Determine the number of moles of $CaCl_2$ in 6.24 g of this sample.

7. The thermite reaction, an explosive reaction between aluminum and iron(III) oxide, requires 12.68 g of iron(III) oxide. Determine the number of iron(III) oxide molecules in this sample.

8. Determine the number of moles of hydrazine (N_2H_4), a common rocket fuel, in 12.63 g of this compound.

Make sure you practice more problems from your chemistry textbook to master this concept since it is essential for your understanding of future topics in this workbook.

Please complete each part of each question before proceeding forward. **Write down your problem solving strategy; not just the answers.**

1.   A student separated the different colored M&M candies in a packet and collected the following data.

Color of candies	Number of candies	Mass (g)
Red	9	45
Orange	12	60
Yellow	10	50
Brown	14	70
Total		

Answer the questions based on the above data. The first two parts are done for you. Note that the problem solving strategy is more important than the final answer.

   a)   What fraction of the candies is red colored?

-    First note that there are 45 candies total (9+12+10+14 = 45 candies)

-    $\text{fraction of red candies} = \dfrac{(\text{number of red candies})}{(\text{total number of candies})} = \dfrac{9}{45}$

   b)   What percentage of the candies is red colored?

-    $\text{percentage of red candies} = \text{fraction of red candies} \times 100\% = \dfrac{9}{45} \times 100\% = 20\%$

   c)   What fraction of the candies is orange?

   d)   What percentage of the candies is orange?

e) What fraction of the candies **by mass** is brown?

- In this problem, you are asked to determine the fraction by mass. First note that the total mass of the candies is 225 g (45g+60g+50g+70g = 225 g)

- Fraction by mass of brown candies = $\dfrac{\text{mass of brown candies}}{\text{total mass of candies}} = \dfrac{14\ g}{225\ g}$

f) What percentage of the candies **by mass** is brown?

- percentageby mass of brown candies =

  fraction by mass of brown candies x100% = $\dfrac{14\ g}{225\ g}x100\% = 6.2\%$

g) What fraction of the candies **by mass** is red?

h) What percentage of the candies **by mass** is red?

Note that:

- When you calculated the fraction of colored candies <u>by mass</u> or the percentage of colored candies <u>by mass</u>, the <u>number of candies was not relevant</u>.

- Because the number of candies is not relevant we can use percentage of colored candies by mass to <u>compare between two different packets of candies</u>.

- By **choosing** to express the distribution of colored candies in the packet <u>by mass</u>, (easy to weigh on a scale), we are <u>avoiding the tedious process of having to count the candies</u>.

The following table shows the percentage by mass of each of the colored candies in the packet of M&M candies. Do these calculations on your own and compare with the answers in the table.

Color of candies	Mass (g)	Fraction by mass	Percentage by mass
Red	45	45g/225g	(45g/225g) x 100% = 20%
Orange	60	60g/225g	(60g/225g) x 100% = 27%
Yellow	50	50g/225g	(50g/225g) x 100% = 22%
Brown	70	70g/225g	(70g/225g) x 100% = 31%
Total	**225**		**100%**

Note:

- The **sum of the percentages by mass** of all the colored candies is 100%. What is the reason for this (if you are working in a group, discuss with your team)? Write down the answer in your own words.

- By using the percentage by mass of each of the colored candies, you can determine which colored candies contribute the highest mass (or the lowest mass) to the packet. In this case, the brown colored candies contribute the highest *percent by mass* and the red candies contribute the lowest *percent by mass* to the M&M packet.

- We can also say that the **percent composition by mass** of the M&M packet is: 20% by mass red candies, 27% by mass orange candies, 22% yellow candies and 31% by mass brown candies.

- The process we used above can be extended to study the **percent compositions by mass of elements present in a given compound.** We choose this approach because in compounds and elements (unlike in packages of candies), it is **impossible to count the atoms/molecules**.

Let's extend what we just learned to a simple molecule.

2. Complete the following table for 1 mole of CH_4. Remember to include units!

Element	Number of moles	mass
C		
H		
Total		

a) What is the fraction **by mass** of C in a mole of CH_4? [review the previous problem if needed].

b) What is the percentage **by mass** of C in a mole of CH_4?

The percentage by mass of C in a mole of CH_4 is called the **percent composition by mass of C in CH_4.**

c) Determine the percent composition by mass of H in CH_4.

As in the case of M&M candies packet, the **sum of the percent composition by mass of each element in a compound must add up to 100%** (or very close to it). Verify this by using your answers for the CH_4 molecule.

When solving the following problems, refer to the M&M candies analogy as often as needed.

3. Determine the:

a) Percent composition by mass of Mg in $MgSO_4$

The first step is to generate a table similar to that of the previous problem.

Element	Number of moles	mass
Mg		
S		
O		
Total		

Now determine the percent composition by mass of each of the elements in $MgSO_4$.

a)   Percent composition by mass of Fe in the pyrite mineral [FeS$_2$]

b)   Percent composition by mass of Cu in the chalcopyrite mineral [CuFeS$_2$]

c) Percent composition by mass of Fe in the chalcopyrite mineral [$CuFeS_2$]

d) Based on the above, if you were to select a mineral ore to extract copper, which mineral (out of pyrite and chalcopyrite) would you choose and why?

e) Percent composition by mass of Cu in the malachite mineral [$Cu_2CO_3(OH)_2$]

4. The primary minerals for extracting sodium are halite (NaCl) and soda ash (Na_2CO_3). Based on the percent by mass of sodium, which mineral would yield the highest amount of sodium?

5. The mineral bornite (also called the peacock mineral due to its iridescent, beautiful colors) with the chemical formula Cu_5FeS_4 and the mineral chalcopyrite with the chemical formula $CuFeS_2$ contain copper and iron. Which of these minerals has the higher percent by mass of copper?

6. Caffeine ($C_8H_{10}N_4O_2$ – found in coffee) and theobromine ($C_7H_8N_4O_2$ – found in chocolate) are two compounds familiar to all college students. Which one of these compounds has the higher percent by mass of nitrogen?

7. Taurine, sometimes known as Tauric acid, is an ingredient in energy drinks. The chemical formula of taurine is $C_2H_7NO_3S$. Calculate the percent by mass of nitrogen in taurine.

The problems above are helpful in learning the concept of percent composition by mass. Be sure to practice more problems from your chemistry textbook to master this concept.

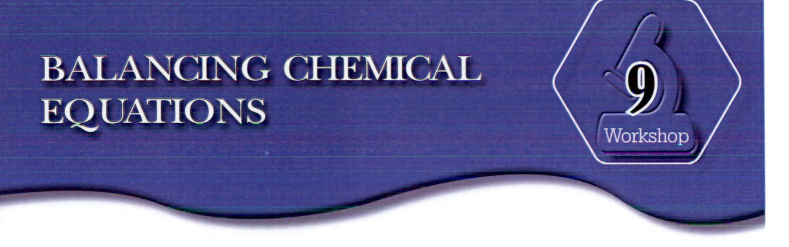

A chemical reaction involves **reactants** and **products**. The reactants react together to form products. Think about baking a cake in your kitchen. The ingredients you mix together (sugar, flour, **eggs**, etc.) are your reactants and the cake is your product. We know that a chemical reaction has occurred because the product (cake) is fundamentally different from the reactants (sugar, flour, eggs, etc.). A chemical reaction involves breaking and forming bonds (i. e. rearrangement of atoms) regardless of whether the reaction is done in your kitchen or in the laboratory.

A chemical reaction is often represented by a **balanced chemical equation**. We use a **balanced** chemical equation because "**matter cannot be created or destroyed**" (also known as the law of conservation of matter) during a chemical reaction. In this workshop, we will learn to balance chemical equations.

When writing chemical equations make sure that:

- The reactants are written on the left hand side
- The products are written on the right hand side
- The reactants/products are written using chemical formulas
- An arrow points from the reactants to the products. The arrow may be replaced by an equal sign.
- The phase of each reactant/product is written in parenthesis after the chemical formula of each species. The solid phase is designated by (s), the liquid phase by (l), the gas phase by (g) and an aqueous solution by (aq).

Example:

$$CH_4 \text{ (g)} + O_2 \text{ (g)} \rightarrow CO_2 \text{ (g)} + H_2O \text{ (g)} \qquad\qquad \text{Equation A}$$

Note that CH_4 and O_2 are reactants and CO_2 and H_2O are products. In this particular reaction, all reactants and products are in the gaseous phase.

Now let us focus on balancing this chemical equation. Remember that we follow the law of conservation of matter when balancing chemical equations. Since "**matter cannot be created or destroyed**", we must make sure that the **number of atoms of <u>each element</u> on the reactant side is equal to the number of atoms of that element on the product side**.

To do so we must keep count of the number atoms of each element on the reactant and product sides.

element	# atoms on the reactant side	# atoms on the product side	Balanced or not
C	1	1	Balanced
H	4	2	Not balanced
O	2	3 (2 O atoms from CO_2 and 1 O atom from H_2O)	Not balanced

Go over Equation A carefully and ensure you are able to fill this table on your own. Note that the number of atoms of each of the elements on the reactant side is NOT equal to the number of atoms of that element on the product side. This means that Equation A is NOT BALANCED.

Note that the number of C atoms is the same on both sides (reactant and product sides) of Equation A. Now let us focus on the H atoms. In order to make the number of H atoms the same on both sides, we can multiply the number of H_2O molecules on the product side by 2.

$$CH_4 \text{ (g)} + O_2 \text{ (g)} \rightarrow CO_2 \text{ (g)} + \mathbf{2}\ H_2O \text{ (g)} \qquad \text{Equation B}$$

Note: When we place the number 2 in front of the H_2O molecule, we are multiplying the entire H_2O molecule by two. This means that the numbers of H atoms and O atoms in the H_2O molecule are now multiplied by 2.

Now count the atoms again to check if Equation B is balanced. First, do this on your own, then check your answer with the following table.

element	# atoms on the reactant side	# atoms on the product side	Balanced or not
C	1	1	Balanced
H	4	4 [2 x H_2O = (2 x H_2) + (2 x O) = **4 x H + 2 x O**]	Balanced
O	2	4 (**2 O** atoms from CO_2 and **2 O** atoms from H_2O as explained above)	Not balanced

Notice that the number of oxygen atoms on the product side increased to 4 when we did the above step. Note that the number of atoms of each element on the reactant side is NOT equal to the number of atoms of that element on the product side. This means that Equation B is NOT BALANCED.

We can make the number of O atoms the same on both sides in Equation B by multiplying the number of O_2 molecules on the reactant side by 2.

$$CH_4 \text{ (g)} + \mathbf{2}\ O_2 \text{ (g)} \rightarrow CO_2 \text{ (g)} + \mathbf{2}\ H_2O \text{ (g)} \qquad \text{Equation C}$$

Now count the atoms again to check if the equation is balanced. First, do this on your own, then check your answer with the following table.

element	# atoms on the reactant side	# atoms on the product side	Balanced or not
C	1	1	Balanced
H	4	4	Balanced
O	4 [2 x O_2 = **4 O**]	4	Balanced

Now note that **the number of atoms of each element on the reactant side is equal to the number of atoms of that element on the product side. Therefore, Equation C is BALANCED.**

The integers in front of each reactant and product in a balanced chemical equation are called **stoichiometric coefficients**. In Equation C, the stoichiometric coefficient for H_2O is 2 and the stoichiometric coefficient for CH_4 is 1. When the stoichiometric coefficient is 1, it is not explicitly written.

- What is the stoichiometric coefficient for O_2 in Equation C?

The following is also a balanced chemical equation for the above reaction.

$$\textbf{2 } CH_4 \text{ (g)} + \textbf{4 } O_2 \text{ (g)} \rightarrow \textbf{2 } CO_2 \text{ (g)} + \textbf{4 } H_2O \text{ (g)} \qquad\qquad \text{Equation D}$$

You can see that Equation D is obtained by multiplying Equation C by 2. We can obtain many different balanced equations by multiplying Equation C by an integer. Try it!

Multiply Equation C by 4: $\qquad \textbf{4 } CH_4 \text{ (g)} + \textbf{8 } O_2 \text{ (g)} \rightarrow \textbf{4 } CO_2 \text{ (g)} + \textbf{8 } H_2O \text{ (g)}$

Multiply Equation C by 7: $\qquad \textbf{7 } CH_4 \text{ (g)} + \textbf{14 } O_2 \text{ (g)} \rightarrow \textbf{7 } CO_2 \text{ (g)} + \textbf{14 } H_2O \text{ (g)}$

Multiply Equation C by 3: $\qquad \textbf{3 } CH_4 \text{ (g)} + \textbf{6 } O_2 \text{ (g)} \rightarrow \textbf{3 } CO_2 \text{ (g)} + \textbf{6 } H_2O \text{ (g)}$

However, when we write a balanced chemical equation we must **use the lowest possible integer values for the stoichiometric coefficients. Therefore, the correct balanced equation for this reaction is Equation C.**

You will learn later that stoichiometric coefficients could be fractions (as shown in the following example). However, **for the purpose of this workshop we will only work with** <u>integer</u> **stoichiometric coefficients.**

$$\tfrac{1}{2} CH_4 \text{ (g)} + O_2 \text{ (g)} \rightarrow \tfrac{1}{2} CO_2 \text{ (g)} + H_2O \text{ (g)}$$

(Balanced with fractions for stoichiometric coefficients)

Note that the chemical equation gives us **qualitative** information by informing us **what** chemicals are reacting and **what** chemicals are being produced. In later workshops you will be able to use balanced chemical equations to obtain **quantitative** information; i.e. **how much** of each chemical is reacting and **how much** of each chemical is produced. **Therefore, the balanced chemical equation is a powerful tool for chemists.**

Let us practice writing balanced chemical equations for the following reactions using the above approach. Be sure to **write down your problem solving strategy, not just the answers.**

1. CaC_2 (s) + H_2O (l) → C_2H_2 (g) + $Ca(OH)_2$ (s)

2. Fe_2O_3 (s) + Al (s) → Al_2O_3 (s) + Fe (l) [thermite reaction]

3. $SiCl_4$ (l) + H_2O (l) → SiO_2 (s) + HCl (aq)

4. CO_2 (g) + H_2O (g) → $C_6H_{12}O_6$ (s) + O_2 (g)

[This is the photosynthesis reaction. $C_6H_{12}O_6$ is glucose]

5.   Zn (s) + HCl (aq) → ZnCl$_2$ (aq) + H$_2$ (g)

6.   Fe (s) + O$_2$ (g) → Fe$_2$O$_3$ (s)   [rusting of iron in the atmosphere]

7. Fe_2O_3 (s) + HNO_3 (aq) → $Fe(NO_3)_3$ (aq) + H_2O (l)

8. NH_3 (g) + O_2 (g) → NO (g) + H_2O (g)

9. $Ca(OH)_2$ (s) + NH_4Cl (g) → $CaCl_2$ (s) + NH_3 (g) + H_2O (l)

10. PCl_5 (s) + H_2O (l) → H_3PO_4 (aq) + HCl (g)

11. CaO (s) + C (s) → CaC_2 (s) + CO_2 (g)

12. MoS_2 (s) + O_2 (g) → MoO_3 (s) + SO_2 (g)   [Mo is molybdenum]

13. $FeCO_3$ (s) + H_2CO_3 (aq) $\rightarrow$ $Fe(HCO_3)_2$ (aq)

14. KO_2 (s) + H_2O (l) $\rightarrow$ KOH (aq) + O_2 (g) + H_2O_2 (l)

15. Fritz Haber, a German Chemist, won the Nobel Prize in chemistry in 1918 for the synthesis of ammonia gas using nitrogen and hydrogen gases as reactants. Write a balanced equation for the synthesis of ammonia gas using the Haber process.

16. $Ba(OH)_2 \cdot 8H_2O$ (s) + NH_4Cl(s) → NH_3(g) + H_2O(l) + $BaCl_2$(s)

[That this is a rare reaction that occurs between two solids]

17. In the laboratory, copper can be made from the mineral malachite [$Cu_2CO_3(OH)_2$]. The first step involves roasting malachite to form copper(II) oxide, carbon dioxide, and water. Write a balanced equation for the first step.

18. The second step involves the reaction of copper(II) oxide with charcoal to produce copper metal, carbon monoxide and carbon dioxide. Write a balanced equation for the second step.

Make sure you practice more problems from your chemistry textbook to master this concept since it is essential for your understanding of future topics in this workbook.

REACTION STOICHIOMETRY IN MOLES

The following is a balanced chemical equation that represents the respiration reaction.

$$C_6H_{12}O_6 \text{ (s)} + 6\,O_2 \text{ (g)} \rightarrow 6\,CO_2 \text{ (g)} + 6\,H_2O \text{ (g)}$$

$C_6H_{12}O_6$ is the chemical formula for glucose. It reacts with oxygen (we breath in oxygen) to make energy for the body (so energy is also a product of this reaction but we only write "matter" in chemical reactions). Other products are water vapor and carbon dioxide (which we exhale).

- What are the stoichiometric coefficients for each of the chemicals in the above reaction?

- What is wrong with writing the above reaction in the following way?

$$2\,C_6H_{12}O_6 \text{ (s)} + 12\,O_2 \text{ (g)} \rightarrow 12\,CO_2 \text{ (g)} + 12\,H_2O \text{ (g)}$$

We use balanced chemical equations to understand chemical reactions **qualitatively (what** are the reactants/products) and **quantitatively (how much** of each reactant/product is needed/formed). Let us use an analogy to help explore this.

The following is a recipe for making one cup of tea.

1 tea bag + 1 cup boiling water + 2 tablespoons milk + 1 teaspoon sugar

So we can write (similar to a *balanced* chemical reaction):

1 tea bag + 1 cup boiling water + 2 tablespoons milk + 1 teaspoon sugar → 1 cup of tea

This is a **qualitative** (**what** is consumed/produced) and **quantitative** (**how much** of each reactant/product is needed/formed) description - very similar to a balance chemical equation.

- If you were to make 5 cups of tea how much sugar would you need?

- How much milk would you need?

Answering the above questions should help you understand that the recipe is actually a **ratio** of the reactants and products. For every cup of tea (product) you need 2 tablespoons of milk (reactant).

We can state this as:

- 2 tablespoons of milk: 1 cup of tea OR

- $\dfrac{2\,tablespoons\,milk}{1\,cup\,of\,tea}$ OR $\dfrac{1\,cup\,of\,tea}{2\,tablespoons\,milk}$

To calculate the amount of milk needed to make 5 cups of tea, we can set up the problem as follows using the above ratio relationship:

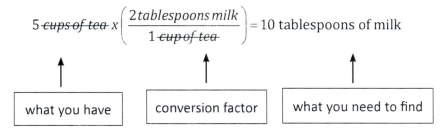

Notice that we used the 2 tablespoons of milk: 1 cup of tea ratio as a conversion factor to solve this problem.

The following ratios are also given in the recipe.

- $\dfrac{1\,tea\,bag}{1\,teaspoons\,sugar}$ OR $\dfrac{1\,teaspoon\,sugar}{1\,tea\,bag}$

- $\dfrac{1\,teaspoons\,sugar}{2\,tablespoons\,milk}$ OR $\dfrac{2\,tablespoons\,milk}{1\,teaspoons\,sugar}$

You must pay careful attention to units in these problems. After all, one teaspoon is not the same as one tablespoon!

Now it is time to apply what we have learned from this exercise to obtain **quantitative** information from a balanced chemical equation. You are given the following balanced reaction.

$$C_6H_{12}O_6\ (s)\ +\ 6\ O_2\ (g)\ \rightarrow 6\ CO_2\ (g)\ +\ 6\ H_2O\ (g)$$

The following quantitative relationships can be obtained from this equation.

- **1 molecule of $C_6H_{12}O_6$ reacts with 6 molecules of O_2 to produce 6 molecules of CO_2 and 6 molecules of H_2O.**

- 1 dozen $C_6H_{12}O_6$ molecules reacts with 6 dozen O_2 molecules to produce 6 dozen CO_2 molecules and 6 dozen H_2O molecules.
- **1 mole of $C_6H_{12}O_6$ reacts with 6 moles of O_2 to produce 6 moles of CO_2 and 6 moles of H_2O.**

It is very important that you understand the above statements. Please go over this carefully (discuss with your teammates if you are working in a group) until you understand them clearly. Note that you cannot obtain quantitative relationships unless you have a **balanced** chemical reaction.

Chemists use the mole relationship (i.e. **1 mole of $C_6H_{12}O_6$ reacts with 6 moles of O_2 to produce 6 moles of CO_2 and 6 moles of H_2O)** far more often than the other two mentioned above.

We can use the balanced chemical equation to say that:

- 1 mole of $C_6H_{12}O_6$: 6 moles CO_2

- $$\frac{1\, mole\, C_6H_{12}O_6}{6\, mole\, CO_2} \quad OR \quad \frac{6\, mole\, CO_2}{1\, mole\, C_6H_{12}O_6}$$

- $$\frac{6\, mole\, O_2}{6\, mole\,\, CO_2} \quad OR\ ?$$

- $$\frac{6\, mole\, H_2O}{1\,\, mole\,\, C_6H_{12}O_6} \quad OR\ ?$$

1. First, balance the following reactions. Then write quantitative mole relationships between the reactants and products. The first one is done for you as an example.

- $CO\ (g) + 2\ H_2\ (g) \rightarrow CH_3OH\ (l)$

    1 mole of CO reacts with 2 moles of H_2 to produce 1 mole of CH_3OH

    $$\frac{1\, mole\, CO}{2\, moles\, of\, H_2} \quad OR \quad \frac{2\, moles\, H_2}{1\, mole\, of\,\, CH_3OH} \quad OR \quad \frac{1\, mole\, CO}{1\, mole\, of\,\, CH_3OH}$$

- $Na\ (s) + \ Cl_2\ (g) \rightarrow \ NaCl\ (s)$

- CH_4 (g) + Cl_2 (g) → CCl_4 (l) + HCl (g)

- Al (s) + H_2SO_4 (aq) → $Al_2(SO_4)_3$ (aq) + H_2 (g)

- C_2H_5OH (l) + O_2 (g) → C_2H_4O (aq) + H_2O (l)

2. You are given the following balanced chemical reaction.

$$4 NH_3 (g) + 5 O_2 (g) \rightarrow 4 NO (g) + 6 H_2O (g)$$

a) Write the mole ratio between NH_3 and NO.

b) If you want to make 3 moles of NO how many moles of NH_3 do you need?

As we have done in the example of making a cup of tea, use the information you have to obtain the information you need. The mole ratio from part (a) above can be used as a conversion factor as shown below.

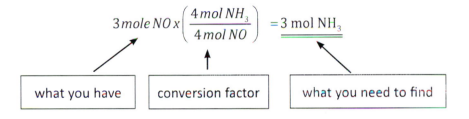

$$3\, mole\, NO\, x \left(\frac{4\, mol\, NH_3}{4\, mol\, NO} \right) = 3\, mol\, NH_3$$

| what you have | conversion factor | what you need to find |

c) If you make 5 moles of NO how many moles of H_2O will be produced?

3. The Haber process involves the synthesis of ammonia gas using nitrogen and hydrogen gases as reactants.

    - Write a balanced equation for the synthesis of ammonia gas using the Haber process.

    - If you want to make 253 moles of ammonia, how many moles of each of the reactants would you need?

4. The thermite reaction, which generates a lot of energy, was used historically to weld railway lines in remote areas where it was difficult to get access to electricity. It is a reaction between two solids as shown below.

- First, balance the reaction. Fe_2O_3 (s) + Al (s) → Al_2O_3 (s) + Fe (l)

- Determine the amount of liquid iron formed if you started with 32 moles of Fe_2O_3

- Determine the amount of aluminum you will need for the above reaction.

- Determine the amount of Al_2O_3 formed during the reaction.

5. Acetylene (C_2H_2), an important industrial chemical and also used in acetylene lamps, is synthesized by the reaction between solid calcium carbide (CaC_2) and water. The products are acetylene gas and solid calcium hydroxide.

- Write a balanced reaction for the synthesis of acetylene.

- Determine the amount of calcium carbide you will need to produce 25.6 moles of acetylene.

- How much of the *byproduct* will be produced? A byproduct is a product that is formed in a chemical reaction in addition to the desired product. In this reaction, the desired product is acetylene. Therefore, calcium hydroxide is the byproduct.

6. The following reaction represents the explosive discharge used by the bombardier beetle to escape from predators. Balance the reaction.

$$C_6H_4(OH)_2 \ (aq) \ + \quad H_2O_2 \ (aq) \rightarrow \quad C_6H_4O_2 \ (aq) \ + \quad H_2O \ (l)$$

- How many moles of each of the reactants is needed to make 12.5 moles of the lethal product?

Be sure to practice more problems from your chemistry textbook to master this concept since it is essential for your understanding of future topics in this workbook.

In the previous workshop, we learned that a balanced chemical reaction could be used to obtain quantitative relationships between reactants and products in units of moles. For example, we can calculate the number of moles of NH_3 needed to make 3 moles of NO as follows.

$$4\ NH_3\ (g)\ +\ 5\ O_2\ (g) \rightarrow\ 4\ NO\ (g)\ +\ 6\ H_2O\ (g)$$

$$3\ \text{moles NO} \times \left(\frac{4\ moles\ NH_3}{4\ \text{moles of NO}} \right) = 3\ \text{moles} NH_3$$

This information is very useful to chemists! One can calculate the quantity of the starting materials needed to make a desired amount of products in any chemical reaction. Conversely one can determine the amounts of products formed from given amounts of reactants. However, it is not practical to "get" 3 moles of NH_3 for this reaction in the laboratory since we do not have a tool to measure out moles (a "mole meter").

However, we do know how to convert moles to mass! Let us use that skill to determine the *mass* of NH_3 needed to make 3 moles of NO.

The molar mass of NH_3 is = 17.034 g/mol (check it!)

Now convert the moles of NH_3 determined above to grams as follows:

$$3\ \text{moles } NH_3 \times \left(\frac{17.034g\ NH_3}{1\ \text{moles of } NH_3} \right) = 51.102\ gNH_3 = 5 \times 10^2\ gNH_3$$

Now we know the *mass* of the reactant (NH_3) needed to make 3 moles of NO *and* we can measure out this amount in the lab using a balance.

In this workshop we will extend what we learned in the previous workshop by using balanced chemical equations to determine masses of reactants and products. Let us try some examples.

Carbon tetrachloride, a useful organic solvent, can be made using the following reaction. Calculate the mass of Cl_2 needed to make 12.5 g of carbon tetrachloride.

$$CH_4\ (g) + 4\ Cl_2\ (g) \rightarrow CCl_4\ (l) + 4\ HCl\ (g)$$

First make sure that the reaction is balanced! Remember that the **balanced** reaction provides relationships between **moles of reactants and products** (not grams). Since we are interested in Cl_2 and CCl_4 we can see from the balanced reaction that the mole ratio is:

$$4 \text{ moles } Cl_2 : 1 \text{ mole } CCl_4 \quad \text{OR} \quad \left(\frac{4 \text{ mol of } Cl_2}{1 \text{mol } CCl_4} \right)$$

We need to make 12.5 g of CCl_4. Since the relationships between reactants and products in a chemical reaction are given in *mole* ratios (not mass), first determine the amount of CCl_4 in moles.

Goal	Calculation
Convert mass of CCl_4 to moles of CCl_4	Molar mass of CCl_4 = 154.01 g/mol (check it!) $$12.5 \text{ g } CCl_4 \times \left(\frac{1 \text{ mol of } CCl_4}{154.01 \text{ g of } CCl_4} \right) = 8.12 \times 10^{-2} \text{ mol } CCl_4$$
Convert moles of CCl_4 to moles of Cl_2	Use the mole ratio between Cl_2 and CCl_4 to obtain the moles of Cl_2 needed. $$8.12 \times 10^{-2} \text{ mol of } CCl_4 \times \left(\frac{4 \text{ mol of } Cl_2}{1 \text{mol } CCl_4} \right) = 3.25 \times 10^{-1} \text{ mol } Cl_2$$
Convert moles of Cl_2 to mass of Cl_2	Convert moles of Cl_2 to mass of Cl_2 by using the molar mass of Cl_2 (70.9 g/mol) (check it!) $$3.25 \times 10^{-1} \text{ mol } Cl_2 \times \left(\frac{70.9 \text{ g } Cl_2}{1 \text{mol } Cl_2} \right) = 23.0 \text{ g } Cl_2$$

Please read the above problem very carefully and make sure you understand it. Work it out on your own (or with your study group) to ensure you have grasped the concept.

Once you get comfortable with this concept, you can set up the problem from beginning to end in one long step as follows.

$$CH_4 \text{ (g)} + 4 \text{ } Cl_2 \text{ (g)} \rightarrow CCl_4 \text{ (l)} + 4 \text{ HCl (g)}$$

Start with **what you have** been given in the problem and work towards **what you need to solve.**

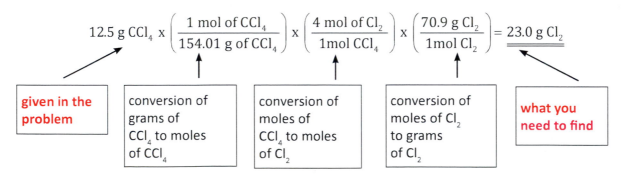

However, there is no requirement to do this all in one step! Take as many steps as you need and work through the problem slowly, making sure you understand each step clearly.

1. Calcium hydroxide reacts with carbon dioxide according to the following reaction. What is the mass of calcium carbonate formed if 30.45 g of calcium hydroxide was used in the reaction.

$$Ca(OH)_2 \text{ (aq)} + CO_2 \text{ (g)} \rightarrow CaCO_3 \text{ (s)} + H_2O \text{ (l)}$$

Check to make sure the reaction is BALANCED before starting any calculations. You can set the problem up the following way. Fill in the blanks in the following table.

Goal	Calculation
Convert mass of $Ca(OH)_2$ to moles of $Ca(OH)_2$	Molar mass of $Ca(OH)_2$ = _____ $30.45 \text{ g } Ca(OH)_2 \times \left(\dfrac{1 \text{ mol } Ca(OH)_2}{\text{___ g of } Ca(OH)_2} \right)$ = _____ mol $Ca(OH)_2$
Convert moles of $Ca(OH)_2$ to moles of $CaCO_3$	Use the mole ratio between $Ca(OH)_2$ and $CaCO_3$ to obtain the moles of $CaCO_3$ formed. _____ mol of $Ca(OH)_2 \times \left(\dfrac{1 \text{ mol of } CaCO_3}{1 \text{mol } Ca(OH)_2} \right)$ = _____ mol $CaCO_3$
Convert moles of $CaCO_3$ to mass of $CaCO_3$	Convert moles of $CaCO_3$ to mass of $CaCO_3$ by using the molar mass of $CaCO_3$ Molar mass of $CaCO_3$ = _____ _____ mol $CaCO_3 \times \left(\dfrac{\text{___ g } CaCO_3}{1 \text{mol } CaCO_3} \right)$ _____ g $CaCO_3$

You can also set this problem up in one step as shown below. With time, you will be able to do these types of problems in one step if you keep practicing. What is important at this point is that you understand each of the steps of the calculation.

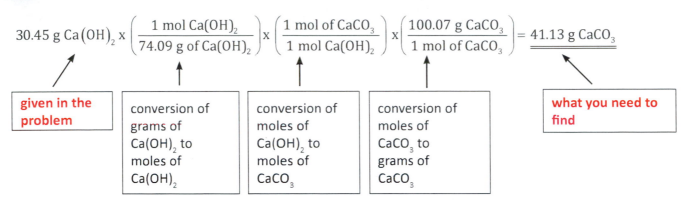

$$30.45 \text{ g Ca(OH)}_2 \times \left(\frac{1 \text{ mol Ca(OH)}_2}{74.09 \text{ g of Ca(OH)}_2} \right) \times \left(\frac{1 \text{ mol of CaCO}_3}{1 \text{ mol Ca(OH)}_2} \right) \times \left(\frac{100.07 \text{ g CaCO}_3}{1 \text{ mol of CaCO}_3} \right) = \underline{\underline{41.13 \text{ g CaCO}_3}}$$

| given in the problem | conversion of grams of Ca(OH)₂ to moles of Ca(OH)₂ | conversion of moles of Ca(OH)₂ to moles of CaCO₃ | conversion of moles of CaCO₃ to grams of CaCO₃ | what you need to find |

With practice, you will be able to do these types of problems in one step. Keep practicing and think through each step! What is important at this point is that you understand each step, regardless of the method you choose to use.

2. Magnesium metal reacts with hydrochloric acid to produce hydrogen gas according to the following reaction. Determine the mass of hydrogen gas produced if 9.58 g of magnesium was used in the reaction.

$$\text{Mg (g)} \quad + \quad \text{HCl (aq)} \rightarrow \quad \text{MgCl}_2 \text{ (aq)} + \quad \text{H}_2 \text{ (g)}$$

3. Ammonium nitrate, a common fertilizer, is produced industrially by the reaction between ammonia and nitric acid. If you desire to produce 245 grams of the fertilizer, how much (in grams) of each of the reactants are needed?

$$HNO_3 \text{ (aq)} + NH_3 \text{ (g)} \rightarrow NH_4NO_3 \text{ (aq)}$$

4. In the automobile, octane gas is combusted to generate energy. The chemical reaction is given below. Determine the mass of CO_2 produced if 12.6 g of octane is combusted (reacted with oxygen).

$$C_8H_{18} \text{ (l)} \quad + \quad O_2 \text{ (g)} \rightarrow \quad CO_2 \text{ (g)} \quad + \quad H_2O \text{ (g)}$$

5. Calcium carbonate, a compound found in the shells of marine organisms, reacts readily with hydrochloric acid to produce carbon dioxide gas according to the following reaction. If 0.395 g of calcium carbonate was used, what is the mass of carbon dioxide produced in the reactions?

$$CaCO_3 \text{ (s)} + HCl \text{ (aq)} \rightarrow CO_2 \text{ (g)} + CaCl_2 \text{ (aq)} + H_2O \text{ (l)}$$

6. When sulfur dioxide, a common industrial pollutant, is released to the atmosphere, it produces sulfuric acid by the following reactions causing acid rain. Determine the mass of sulfuric acid produced from 23.2 g of SO_2.

$$2SO_2 (g) + O_2 (g) \rightarrow 2SO_3 (g) \qquad \text{Reaction 1}$$

$$SO_3 (g) + H_2O (l) \rightarrow H_2SO_4 (aq) \qquad \text{Reaction 2}$$

To solve this problem take one reaction at a time. Use reaction 1 to determine the amount of SO_3 formed and then reaction 2 to determine the amount of H_2SO_4 produced.

7. Smelting of metal ores is a common method for extracting metals. The following reaction represent the smelting of galena (PbS), the most common lead ore, to make PbO and SO_2 (PbO is subsequently reduced to produce the metal Pb). If you started with 55.3 g of galena, how many grams of PbO is produced?

$$PbS \text{ (s)} \ + \ O_2 \text{ (g)} \ \rightarrow \ PbO \text{ (s)} \ + \ SO_2 \text{ (g)}$$

8. Fritz Haber, a German chemist, won the Nobel Prize in chemistry in 1918 "*for the synthesis of ammonia from its elements*" using the following reaction. Determine the masses of hydrogen and nitrogen gases needed to synthesize 750 kg of ammonia gas. [hint: start with a balanced reaction]

$$H_2 (g) \quad + \quad N_2 (g) \quad \rightarrow \quad NH_3 (g)$$

Make sure you practice more problems from your chemistry textbook to master this concept since it is essential for your understanding of future topics in this workbook.

LIMITING REACTANTS

You want to make chocolate chip cookies. The supplies (reactants) available in the kitchen are given on the left hand side. Your recipe is on the right hand side.

Available in the kitchen	Recipe for 25 chocolate chip cookies
2 lb Butter	½ lb Butter
5 lb Sugar	1 cup Sugar
3 lb Brown sugar	½ cup Brown sugar
1 dozen Eggs	2 Eggs
5 lb Flour	3 cups Flour
1 lb Salt	1 teaspoon Salt
250 mL Vanilla	1 tablespoon Vanilla
¼ lb Baking soda	2 teaspoons Baking soda
1 package (16 oz) Chocolate chips	16 oz Chocolate chips

- Given the above information, what is the maximum number of chocolate chip cookies (**product**) you can make?

- Describe in words, the process you went through to obtain the answer to the question above.

- If you want to make 50 cookies (**product**), what would you need to do?

Note that in the above exercise, the number of cookies (**product**) you were able to make was limited by the amount of one of the ingredients (**reactant**) available in the kitchen. Since this reactant *limits your ability to make the desired amount of product*, it is called the **limiting reactant**.

- What was the limiting reactant in the above exercise?

Note that the other ingredients (reactants) are in excess (therefore called **excess reactants**). Some quantity of each of the excess reactants will be left over after 25 cookies are made.

- Determine the amount of one of the excess reactants that would be left over after you make 25 cookies.

In the chemistry laboratory and in chemical industry one reactant (ingredient), often due to its high cost, limits the amount of desired product that can be produced. This ingredient is called the **limiting reactant.** Since the **limiting reactant determines how much of the desired product can be made**, it is very important to **identify the limiting reactant** so one can **predict** the amount of the desired product.

- Why is it important to predict the amount of product?

Let us try an example to illustrate this point.

The following balanced reaction is used to make ammonia for the fertilizer and chemical explosives industries. The German chemist, Fritz Haber, won the Nobel Prize in Chemistry in 1918 for discovering a method to make ammonia using this reaction (you can search the web for more information about this prize).

If 16.81 g of N_2 is combined with 2.428 g of H_2 what mass of the desired product is formed?

$$N_2 (g) + 3 H_2 (g) \rightarrow 2 NH_3 (g)$$

Note that this problem is not very different from the chocolate chip cookie problem we studied above!

- We have a specified amount of each of the reactants (similar to the amounts of ingredients available in the kitchen)

- We have the balanced chemical equation (similar to a recipe used in the kitchen) that determines the quantity relationships between the reactants and the desired product

Since we know that the quantity relationships between the reactants and the products in a chemical reaction are in units of moles (NOT GRAMS), first convert each of the given amounts of reactants to moles.

Reactants in grams	conversion to moles	Reactants in moles
16.81 g of N_2	$16.81 \text{ g } N_2 \text{ x } \left[\dfrac{1 \text{ mole } N_2}{28.014 \text{ g } N_2} \right] =$	**0.6001 moles of N_2**
2.428 g of H_2	$2.428 \text{ g of } H_2 \text{ x } \left[\dfrac{1 \text{ mole } H_2}{2.016 \text{ g } H_2} \right] =$	**1.2043 moles of H_2**

Recall that the chocolate chip recipe tells us the **ratio** of the ingredients needed to make 25 chocolate chip cookies. For example:

½ lb butter : 1 cup sugar: 2 eggs: 3 cups flour: 16 oz chocolate chips:25 chocolate chip cookies etc.

Similarly, the balanced chemical reaction tells us the **ratio** of the reactants needed to make the desired products in **mole** units (as you learned in a previous workshop).

1 mole of N_2: 3 moles of H_2: 2 mole of NH_3

Now, just as we did in the chocolate chip recipe, we must find out if we have a **limiting reactant,** and if so, **which of the reactants is the limiting reactant.**

Step 1: You may start with any one of the reactants. So let us start with N_2. Then calculate how much H_2 is needed to consume all of the N_2.

$$0.6001 \text{ moles of } N_2 \text{ x } \left[\frac{3 \text{ mol } H_2}{1 \text{ mol of } N_2} \right] = 1.8003 \text{ mol } H_2$$

Now write the above answer using a complete sentence: **1.8003 moles H_2 will be needed to consume all of the N_2 that is available.**

Do we have 1.8003 moles H_2?

- If the answer is yes, then you are not limited by the amount of H_2 (i.e. H_2 is not the limiting reactant)
- If the answer is no, then you are limited by the amount of H_2 (i.e. H_2 is the limiting reactant)

Given the above information what is the limiting reactant?

Step 2: Now repeat this calculation, but start with the given amount of H_2 to determine the limiting reactant.

$$1.2043 \text{ moles of } H_2 \text{ x } \left| \frac{1 \text{ mol } N_2}{3 \text{ mol of } H_2} \right| = 0.4014 \text{ mol } N_2$$

Now write the above answer using a complete sentence: **0.4014 moles N_2 will be needed to consume all of the H_2 that is available.**

Do we have 0.4014 moles N_2?

- If the answer is yes, then you are not limited by the amount of N_2 (i.e. N_2 is not the limiting reactant)
- If the answer is no, then you are limited by the amount of N_2 (i.e. N_2 is the limiting reactant)

Given the above information what is the limiting reactant?

Evaluate: Note that regardless of which reactant you started with (Step 1 or Step 2 above), the limiting reactant is the same – in this case it is H_2 (because we do not have enough H_2 to react with all of the available N_2).

Now that the limiting reactant has been determined, recognize that the amount of products will depend on the limiting reactant (recall that when baking cookies, the number of cookies you could make depended on the amount of chocolate chips, the limiting reactant).

When the reaction is over, all of the limiting reactant (H_2) will be consumed, some of the excess reactant (N_2) will be left over, and some product (NH_3) will be formed.

Read the above statements carefully several times and make sure you comprehend it. Compare it to the chocolate chip cookie problem above. If you are working with a group, discuss this statement with your team.

- Calculate the amount of product formed (yield).

To do this, **one must now start with the amount of the limiting reactant.**

$$1.2043 \text{ moles of } H_2 \text{ x } \frac{2 \text{ mole } NH_3}{3 \text{ mole of } H_2} \text{ x } \frac{17.034 \text{ g } NH_3}{1 \text{ mole } NH_3} = 13.68 \text{ g } NH_3 \text{ will be formed}$$

The amount of product is often called the "yield." In this problem our yield is 13.68 g NH_3. Since this yield was determined using calculations (theory) and not in the lab, this is called the **theoretical yield**.

Therefore the theoretical yield of $NH_3 = \underline{13.68 \text{ g}}$

- How much of the excess reagent is left over?

Recall that the excess reagent is N_2. First calculate how much N_2 is consumed during the reaction.

$$1.2043 \text{ moles of } H_2 \ \text{x} \left| \frac{1 \text{ mole } N_2}{3 \text{ mole of } H_2} \right| \text{x} \left| \frac{28.014 \text{ g } N_2}{1 \text{ mole } N_2} \right| = 11.25 \text{ g } N_2 \text{ will be consumed}$$

Mass of left over N_2 = mass of N_2 available – mass of N_2 consumed

= (16.81 – 11.25) g = 5.560 g

- A scientist did this experiment in the laboratory and was able to make 9.23 g of NH_3. What is the percent yield?

The amount of NH_3 made in the laboratory is called the **actual yield** or the **experimental yield**. Percent yield is defined as follows. Use this definition to calculate the percent yield of ammonia in the above problem.

$$\text{Percent yield} = \left| \frac{\text{actual yield}}{\text{theoretical yield}} \right| x\,100\%$$

$$= \left| \frac{9.23 \text{ g } NH_3}{13.68 \text{ g } NH_3} \right| x\,100\% = 67.5\%$$

The actual yield is always less than the theoretical yield. This is because it is impossible to not lose some of the product during the chemical synthesis process (because chemicals stick to walls of the containers or reactors, escape from valves, dissolve in solvents, etc.)

Read the above problem carefully. Discuss with group members if you are working with a team. Then try the following problems. The first one is done as an example.

1. Iron and chorine gas are reacted to make solid iron(III) chloride. How much of this product is formed if 10.0 g of each of the reactants is used in the reaction? (Hint: start with the balanced equation of the reaction)

$$2 \text{ Fe (s)} + 3 \text{ Cl}_2 \text{ (g)} \rightarrow 2 \text{ FeCl}_3 \text{ (s)}$$

Since the reaction (recipe) provides information in mole relationships, convert all data given to moles first.

$$10.0 \text{ g Fe x} \left| \frac{1 \text{ mol Fe}}{55.85 \text{ g Fe}} \right| = 0.179 \text{ mol Fe}$$

$$10.0 \text{ g Cl}_2 \ \text{x} \left| \frac{1 \text{ mol } Cl_2}{70.90 \text{ g } Cl_2} \right| = 0.141 \text{ mol Cl}_2$$

Now determine the limiting reactant. You can start with either reactant. Let us start with Fe and determine the amount of Cl_2 needed to react with all the Fe that is available.

$$0.179 \text{ mol Fe} \times \left| \frac{3 \text{ mol } Cl_2}{2 \text{ mol Fe}} \right| = 0.269 \text{ mol } Cl_2 \text{ needed to react with all the Fe that is available.}$$

Since we have only 0.141 moles of Cl_2, we can see that Cl_2 is the limiting reactant. Now that we have determined the limiting reactant, recall that the amount of product ($FeCl_3$) formed is dependent on the amount of the limiting reactant.

$$0.141 \text{ mol } Cl_2 \times \left| \frac{2 \text{ mol } FeCl_3}{3 \text{ mol } Cl_2} \right| = 0.0940 \text{ mol Fe } Cl_3$$

$$0.0940 \text{ mol } FeCl_3 \times \left| \frac{162.2 \text{ g } FeCl_3}{1 \text{ mol } FeCl_3} \right| = 15.2 \text{ g } FeCl_3$$

2. Sodium sulfide is used in the leather industry to remove hair from hides. It is made by the following reaction. If you mix 15 g of sodium sulfate and 7.5 g of carbon, what mass of sodium sulfide is produced?

$$Na_2SO_4 \text{ (s)} + \quad C \text{ (s)} \rightarrow \quad Na_2S \text{ (s)} + \quad CO \text{ (g)}$$

3. Ammonia gas can be produced by the following reaction. If 112 g of calcium oxide and 224 g of ammonium chloride are mixed, what mass of ammonia is produced?

$$NH_4Cl \text{ (s)} + CaO \text{ (s)} \rightarrow NH_3 \text{ (g)} + H_2O \text{ (g)} + CaCl_2 \text{ (s)}$$

4. Aspirin, $[C_6H_4(OCOCH_3)CO_2H]$, is produced by the reaction between salicylic acid, $[C_6H_4(OH)CO_2H]$, and acetic anhydride, $[(CH_3CO)_2O]$, according to the following balanced reaction.

$$C_6H_4(OH)CO_2H \text{ (s)} + (CH_3CO)_2O \text{ (l)} \rightarrow C_6H_4(OCOCH_3)CO_2H \text{ (s)} + CH_3COOH \text{ (l)}$$

If you mix 100. g of each of the reactants, what mass of aspirin is produced? Determine the left over amount of the excess reactant.

5. The photosynthesis reaction produces solid glucose ($C_6H_{12}O_6$) and oxygen gas from atmospheric carbon dioxide gas and water. If 10.0 g of each of the reactants is used, calculate the theoretical yield of glucose. If 5.28 g of glucose was actually formed, what is the percent yield?

6. Sodium metal reacts explosively with chlorine gas to make sodium chloride (table salt). Write a balanced reaction and determine the mass of sodium chloride formed if 12.3 g of sodium is mixed with 10.8 g of chlorine. What is the excess reactant and how much of it is left after the reaction is complete?

7. Silicon dioxide (sand) reacts with carbon to make solid silicon carbide (SiC) and carbon monoxide. Determine the mass of silicon carbide that would be formed by mixing 25.3 g of silicon dioxide and 12.5 g of carbon. Determine how much of the excess reagent will be left after the reaction is complete. If the reaction produced 10.4 g of silicon carbide in the laboratory, what is the percent yield?

The problems above are helpful in learning the concept of limiting reactants. Be sure to practice more problems from your chemistry textbook to master this concept.

Write the definition for "homogenous mixture." What is another name for homogenous mixture?

A solution is made of at least two components:

- **Solvent**: the component that makes up the largest portion of the solution
- **Solute**: components that makes up smaller portions of the mixture relative to the solvent

Water is a very common solvent in the lab and in nature. It is abundant and is very good at dissolving things. A solution that is made using water as the solvent is called an **aqueous solution.** Water is such a common solvent that unless otherwise specified, we assume that water is the solvent in a solution.

- What kind of solutes dissolve in water?

- A solution can have only one solvent but it can have many solutes. Think about solutions you know that have more than one solvent. Then complete the following table.

Solution	Solvent	Solute(s)
salt solution		
drinking alcohol		
brass		
atmospheric air		

- We have all made solutions in the kitchen. Think about making a glass of apple juice for breakfast starting with concentrated apple juice (from a can in the freezer). Make two different glasses of apple juice, the first by adding two tablespoons of the concentrated juice and the second by adding three tablespoons of the concentrated juice. Now add enough water to fill the glasses and stir them to make two clear glasses of juice you can drink with breakfast.

- Which of the above two glasses is more concentrated?

    Note that we used the word "concentrated" to describe the glass that was made by adding three table-spoons of the concentrated juice. In comparison to this glass, the one made with two tablespoons of the concentrated juice would be "dilute."

Although we can be somewhat lax when making a glass of apple juice starting from the concentrate in our kitchens, chemists must be very precise when preparing solutions. Imagine a chemist making up a saline solution or a chemotherapy drug that is injected into a patient's IV fluid. These solutions must be very carefully prepared since any changes in concentrations could hurt the patient. Many solutions are made by chemists every day using careful measurements. This workshop is about learning how to determine concentrations of solutions accurately.

- Chemists quantify the **concentrations** of solutions in multiple ways. The most common, and the one preferred in first-year chemistry, is "molarity."

$$molarity = \frac{number\ of\ moles\ of\ solute}{total\ volume\ of\ solution\ in\ liters} \quad : \quad units = \frac{moles}{liters} = mol/L = mol\ L^{-1}$$

Molarity is given the symbol M and M is often written in place of mol L^{-1}. Other ways in which concentrations of solutions are expressed include: g of solute/L of solution, parts per million (ppm), molality, and parts per billion (ppb).

One common mistake students make is using the volume of solvent instead of the volume of solution in the denominator when calculating molarity. Be careful not to make this mistake!

Let us apply this knowledge to a problem.

1. If you used 0.245 moles of NaCl and prepared a solution by adding enough water to make the total volume of 1.50 L, what is the concentration of your salt solution?

$$Remember\ that\ molarity = \frac{number\ of\ moles\ of\ solute}{total\ volume\ of\ solution\ in\ liters}$$

The number of moles of solute (NaCl) and the volume of the solution are given to us.

$$Molarity = \frac{0.245\ mol\ NaCl}{1.50\ L\ solution} = 0.233\ mol\ L^{-1} = 0.233\ M$$

2. If you used 0.369 moles of KCl and prepared a solution by adding enough water to make the total volume of 150.00 mL, what is the concentration of your solution? [hint: note that the volume of the solution is given in mL. First, you must convert this to liter units before calculating the molarity. Why is that?]

3. If the molarity of a 25.00 mL NaOH solution is 1.34 M, how many moles of NaOH must be present in it? [hint: this problem may appear different, but if you use the definition for molarity you will be able to solve it]

If we know the concentrations of a solution accurately, such a solution is called a **standard solution**. Preparing standard solutions often require using volumetric flasks and pipettes.

4. If you prepared a $Cu(NO_3)_2$ solution in a 100 mL volumetric flask by weighing out 2.4530 g of $Cu(NO_3)_2$ into the flask and adding enough DI water to make up to the mark:

    a) What is the volume of the above solution?

    b) How many moles of $Cu(NO_3)_2$ have you added to the flask?

    c) Calculate the concentration of the solution in mol L^{-1} units

5. If you have 25.34 mL of 1.25 M standard solution of $Fe(NO_3)_3$

    a) What is the volume of the solution in liters?

    b) How many moles of $Fe(NO_3)_3$ are present in this solution?

    c) How many moles of iron ions are present in this solution?

    d) What mass of $Fe(NO_3)_3$ must you weigh out in order to prepare this solution?

6. A titration requires you to prepare 50.00 mL of 1.25 M NaOH solution. Determine the mass of solute needed for this preparation.

• What glassware and other equipment would you use in the laboratory when preparing this solution?

7. You want to prepare 100.00 mL of 0.325 M solution of NH_4NO_3. How would you prepare this solution in the laboratory?

8. Silver nitrate [$AgNO_3$] is used in the chemical industry as a precursor to other silver compounds since it is a relative inexpensive salt that dissolves readily in water. How many grams of solid silver nitrate is required to prepare 100.00 mL of 0.101 M silver nitrate solution? How would you prepare this solution in the laboratory?

It is critical to practice this newly acquired skill by doing problems from your chemistry textbook.

We are familiar with purchasing concentrated frozen juice cans from the grocery store and preparing a glass of juice by adding water to the concentrate. This process is called **diluting**.

Juice is sold in concentrated cans for multiple reasons.

1. It costs less to ship and store small concentrated cans compared to the ready to drink bottles of juice (bulky and therefore requires more space).

2. Consumers can choose their desired concentration when preparing the diluted juice. Ready to drink bottles offer only one concentration to the consumer.

Liquid chemicals are also shipped in concentrated form (not frozen however) to reduce the cost of storage and shipping. Your university likely has a "lab store" that purchases and stores these concentrated liquid chemicals, called **stock solutions**. If you need a diluted solution, it is prepared starting from the stock solution by a process called **quantitative dilution,** much like the scenario of preparing a diluted glass of juice from the concentrated can.

For example, acids and bases (nitric acid, hydrochloric acid, sulfuric acid and ammonia solution) are shipped in concentrated form. Then they are **quantitatively diluted** as desired for experiments. It is economical to purchase one concentrated acid solution and prepare as many diluted solutions from it as needed from it than to purchase every possible diluted solution of a given acid.

Let us take a closer look at preparing a glass of juice starting from the concentrate. You can prepare a glass of juice by mixing 2 tablespoons of the concentrate and one cup of water. You can prepare a second glass of juice by mixing 4 tablespoons of concentrate and one cup of water. The second glass will be sweeter and probably tastier! This is the beauty of the concentrated frozen concentrate! A diluted solution of desired concentration can be made from it.

Now consider the following scenario.

John mixes 2 tablespoons of the concentrate and one cup of water to prepare his juice.

Susan mixes 2 tablespoons of the concentrate and two cups of water to prepare her juice.

- Who has the more concentrated juice (and why)?

Both John and Susan have the same amount of concentrate (solute) in their drinks. The only difference is the amount of water (solvent) they added. Unlike in the kitchen, we have to be very careful with our measurements when we prepare dilute solutions starting from a stock solution in the laboratory, since our experimental results depend on it. Hospitals often prepare dilute solutions from stock solutions for their patients. In this case, someone's life could depend on the hospital staff's ability to correctly prepare a quantitatively diluted solution!

You want 100.00 mL of 1.523 M HCl solution for a titration. You lab stores has a stock solution of HCl that is labeled 11.65 M. Let us figure out how to prepare this desired dilute solution by **quantitative dilution**.

It is clear that you need to take some amount (volume) of HCl from the stock solution and add some amount of water to it to prepare the desired diluted solution. The questions we have to answer are:

- how much of the concentrated HCl do you take from the stock solution?

- how much water do you add so that at the end you have 100.00 mL of 1.523 M HCl solution?

This problem requires us to remember that **matter cannot be created or destroyed by dilution**! If we take a certain amount of HCl (solute) from the stock solution, that amount of HCl (solute) will remain in the diluted solution even after we add water to it. For an analogy think about the juice that John and Susan made. They each have 2 tablespoons of the concentrate (solute) in their juice even after dilution.

We can say that **number of teaspoons of solute (frozen juice concentrate) in the diluted juice = number of teaspoons of solute taken from the concentrate**

In chemistry, we measure matter in moles, not teaspoons. The guiding principle for quantitative dilution is:

The number of moles of solute in the dilute solution = number of moles of solute taken from the stock solution

Can we figure out the number of moles of solute in the dilute solution or the number of moles of solute taken from the stock solution? To answer this question let us gather the information we have about each of these solutions.

Table 14:1

	Information on the dilute solution	Information on the stock solution
volume	100.00 mL	?
concentration	1.523 M	11.65 M

It is clear from the above table that we have more information about the dilute solution than the stock solution. So let us see if we can figure out the number of moles of solute in the dilute solution.

$$\text{Recall that molarity} = \frac{number\ of\ moles\ of\ solute}{volume\ of\ solution\ in\ liters}$$

We know the molarity and the volume of the dilute solution (100.00 mL = 1.0000×10^{-1} L). Therefore, we CAN use this information to calculate the moles of solute in this dilute solution.

Number of moles of solute in the dilute solution = molarity of dilute solution x volume of dilute solution in liters

Number of moles of solute in the dilute solution =   1.523 M x1.0000x10^{-1} L = 0.1523 mol

Since the guiding principle for quantitative dilution is:

The number of moles of solute in the dilute solution = number of moles of solute taken from the stock solution

Number of moles of solute in the dilute solution = 0.1523 mol = number of moles of solute taken from the stock solution

Using the definition of molarity we can say that:

Number of moles of solute taken from the stock solution = molarity of the stock solution x volume taken from the stock solution in liters

0.1523 mol = 11.65 M x volume taken from the stock solution in liters

Volume taken from the stock solution in liters $= \dfrac{0.1523 \text{ mol}}{11.65 M} = 1.307 \times 10^{-2} \text{L}$

$= 13.07 \text{ mL}$

Check to make sure you are getting the units right!

Now we can complete Table 14.1

	Information on the dilute solution	Information on the stock solution
volume	100.00 mL	**13.07 mL**
concentration	1.523 M	11.65 M
moles of solute	**0.1523 mol**	**0.1523 mol**

Now that the calculations are done, how do we prepare **100.00 mL of 1.523 M HCl** solution in the laboratory? Since this is a quantitative dilution, we must use precise glassware and lab equipment for this preparation (we cannot use approximate tools such as teaspoons or cups).

We use a burette to obtain **13.07 mL of the HCl stock solution** and we place it in a **100 mL** volumetric flask (it reads to 100.00 mL). Then we **add enough DI water** to make up to the 100 mL mark. We have now prepared **100.00 mL of 1.523 M HCl** solution.

Note: We did not add 100 mL of DI water! We added **enough water so that the final volume of the diluted solution is 100.00 mL**. It is very important to understand this distinction. Please read over carefully and make sure you understand not only the calculations but also the process of preparing the diluted solution. The success of your experiment (or someone's life) will depend on it!

- Your lab partner needs 50.00 mL of 3.237 M HCl solution for a titration. Can you use the same HCl stock solution you used to prepare this solution? If so why? If not, why?

1. How would your lab partner prepare 50.00 mL of 3.237 M HCl solution starting from the stock solution of HCl that is labeled 11.65 M?

The guiding principle for quantitative dilution is:

The number of moles of solute in the dilute solution = number of moles of solute taken from the stock solution

Can we figure out the number of moles of solute in the dilute solution or the number of moles of solute taken from the concentrated solution? To answer this question, completer Table 14.2

Table 14:2

	Information on the dilute solution	Information on the stock solution
volume	50.00 mL	?
concentration	3.237 M	11.65 M

It is clear from the above table that we have more information about the dilute solution than the stock solution.

- Using molarity $= \dfrac{number\ of\ moles\ of\ solute}{volume\ of\ solution\ in\ liters}$ calculate the number of moles of solute in the dilute solution.

- Since the guiding principle for quantitative dilution is:

The number of moles of solute in the dilute solution = number of moles of solute taken from the stock solution, calculate the number of moles of solute taken from the stock solution.

- Using the definition of molarity, calculate the volume taken from the stock solution in liters. You may want to convert this volume into mL for convenience.

- Now complete Table 14.2

Table 14:2

	Information on the dilute solution	Information on the stock solution
volume	50.00 mL	
concentration	3.237 M	11.65 M
moles of solute		

- Now write a description of how your lab partner would prepare 50.00 mL of 3.237 M HCl solution using the stock solution of HCl. Be sure to refer to the needed glassware and lab equipment.

2. In the laboratory, you are provided with a 10.3452 M KNO_3 stock solution. Use this to prepare:

- 25.00 mL of 1.5308 M KNO_3 solution

- 10.00 mL of 2.2846 M KNO_3 solution

- 100.00 mL of 2.5467 M KNO_3 solution

- 50.00 mL of 3.6792 M KNO_3 solution

- 250.00 mL of 6.3942 M KNO_3 solution

3. You find a stock solution of $KMnO_4$ in a lab room. The label is damaged and it is difficult to read the concentration. A student's lab notebook found nearby indicates that she used this stock solution to prepare 25.00 mL of 2.002 M diluted solution. The detailed information in the student's notebook states that 11.34 mL of the stock solution was taken using a burette to prepare the dilute solution. Use this information to determine the concentration of the stock solution. [hint: set up a table similar to Table 14.1 to start the solution].

4. How would you prepare 500.00 mL of 2.345 M HNO_3, starting from a stock solution of 18.23 M?

It is critical to practice this newly acquired skill by doing problems from your chemistry textbook.

Appendix

References:

(1) Felder, R. M. Stoichiometry without Tears. *Chem. Eng. Educ.* **1990**, *24* (4), 188–196.

(2) Cooper, C. I.; Pearson, P. T. A Genetically Optimized Predictive System for Success in General Chemistry Using a Diagnostic Algebra Test. *J. Sci. Educ. Technol.* **2012**, *21* (1), 197–205.

(3) Rowe, M. B. Getting Chemistry off the Killer Course List. *J. Chem. Educ.* **1983**, *60* (11), 954.

(4) Bent, H. A. Should the Mole Concept Be X-Rated? *J. Chem. Educ.* **1985**, *62* (1), 59.

(5) Copley, G. N. The Mole in Quantitative Chemistry. *J. Chem. Educ.* **1961**, *38* (11), 551.

(6) Milio, F. A. Mole Concept and Limiting Reagent in the Laboratory. *J. Chem. Educ.* **1971**, *48* (3), 155.

(7) Duncan, I. M.; Johnstone, A. H. The Mole Concept. *Educ. Chem.* **1973**, *10* (6), 213–214.

(8) Kolb, D. The Mole. *J. Chem. Educ.* **1978**, *55* (11), 728.

(9) Brown, B. S. A Mole Mnemonic. *J. Chem. Educ.* **1991**, *68* (12), 1039.

(10) Krishnan, S. R.; Howe, A. C. The Mole Concept: Developing an Instrument To Assess Conceptual Understanding. *J. Chem. Educ.* **1994**, *71* (8), 653.

(11) Dominic, S. What's a Mole For? *J. Chem. Educ.* **1996**, *73* (4), 309.

(12) Krieger, C. R. Stoogiometry: A Cognitive Approach to Teaching Stoichiometry. *J. Chem. Educ.* **1997**, *74* (3), 306–null.

(13) Wakeley, D. M.; de Grys, H. Developing an Intuitive Approach to Moles. *J. Chem. Educ.* **2000**, *77* (8), 1007.

(14) Jensen, W. B. The Origin of the Mole Concept. *J. Chem. Educ.* **2004**, *81* (10), 1409.

(15) Longo, K. J. Using a Socratic Dialogue To Teach the Mole Concept to Adult Learners. *J. Chem. Educ.* **2007**, *84* (8), 1285.

(16) Fang, S.-C.; Hart, C.; Clarke, D. Unpacking the Meaning of the Mole Concept for Secondary School Teachers and Students. *J. Chem. Educ.* **2014**, *91* (3), 351–356.

(17) Gulacar, O.; Eilks, I.; Bowman, C. R. Differences in General Cognitive Abilities and Domain-Specific Skills of Higher- and Lower-Achieving Students in Stoichiometry. *J. Chem. Educ.* **2014**, *91* (7), 961–968.

(18) Schmidt, H. Secondary School Students' Strategies in Stoichiometry. *Int. J. Sci. Educ.* **1990**, *12* (4), 457–471.

(19) Gable, D.; Sherwood, R. Analysing Difficulties with Moleconcept Tasks by Using Familiar Analog Task. *J Res Sci Teach* **1984**, *21*, 843–851.

(20) Frazer, M. J.; Servant, D. Aspects of Stoichiometry Titration Calculations. *Educ. Chem.* **1986**, *23* (2), 54–56.

(21) Dahsah, C.; Coll, R. K. Thai Grade 10 and 11 Students' Conceptual Understanding and Ability to Solve Stoichiometry Problems. *Res. Sci. Technol. Educ.* **2007**, *25* (2), 227–241.

(22) Wagner, E. P. A Study Comparing the Efficacy of a Mole Ratio Flow Chart to Dimensional Analysis for Teaching Reaction Stoichiometry. *Sch. Sci. Math.* **2001**, *101* (1), 10–22.

(23) Poole, R. L. Teaching Stoichiometry: A Two Cycle Approach. *J. Chem. Educ.* **1989**, *66* (1), 57.

(24) Davidowitz, B.; Chittleborough, G.; Murray, E. Student-Generated Submicro Diagrams: A Useful Tool for Teaching and Learning Chemical Equations and Stoichiometry. *Chem. Educ. Res. Pract.* **2010**, *11* (3), 154–164.

(25) Üce, M. Teaching the Mole Concept Using a Conceptual Change Method at College Level. *Education* **2009**, *129* (4), 683–691.

(26) Ozcan Gulacar, T. L. O. A Novel Code System for Revealing Sources of student's Difficulties with Stoichiometry. *Chem. Educ. Res. Pract.* **2013**, *14* (4), 507-515.

(27) SCHMIDT, H.-J.; JIGNÉUS, C. Students Strategies in Solving Algorithmic Stoichiometry Problems. *Chem. Educ. Res. Pract.* **2003**, *4* (3), 305–317.

(28) Schmidt, H.-J. Stoichiometric Problem Solving in High School Chemistry. *Int. J. Sci. Educ.* **1994**, *16* (2), 191–200.

(29) Furio, C.; Azcona, R.; Guisasola, J. The Learning and Teaching of the Concepts "amount of Substance"and "mole": A Review of the Literature. *Chem. Educ. Res. Pract.* **2002**, *3* (3), 277–292.

(30) Arasasingham, R. D.; Taagepera, M.; Potter, F.; Lonjers, S. Using Knowledge Space Theory To Assess Student Understanding of Stoichiometry. *J. Chem. Educ.* **2004**, *81* (10), 1517.

(31) Dori, Y. J.; Hameiri, M. Multidimensional Analysis System for Quantitative Chemistry Problems: Symbol, Macro, Micro, and Process Aspects. *J. Res. Sci. Teach.* **2003**, *40* (3), 278–302.

(32) Griffiths, A. K. A Critical Analysis and Synthesis of Research on Students' Chemistry Misconceptions. *Probl. Solving Misconceptions Chem. Phys.* **1994**, 70–99.

(33) de Astudillo, L. R.; Niaz, M. Reasoning Strategies Used by Students to Solve Stoichiometry Problems and Its Relationship to Alternative Conceptions, Prior Knowledge, and Cognitive Variables. *J. Sci. Educ. Technol.* **1996**, *5* (2), 131–140.

(34) Schmidt, H.-J. An Alternate Path to Stoichiometric Problem Solving. *Res. Sci. Educ.* **1997**, *27* (2), 237–249.

(35) Niaz, M. Cognitive Conflict as a Teaching Strategy in Solving Chemistry Problems: A Dialectic–constructivist Perspective. *J. Res. Sci. Teach.* **1995**, *32* (9), 959–970.

(36) Staver, J. R.; Lumpe, A. T. A Content Analysis of the Presentation of the Mole Concept in Chemistry Textbooks. *J. Res. Sci. Teach.* **1993**, *30* (4), 321–337.

(37) Hafsah, T.; Rosnani, H.; Zurida, I.; Kamaruzaman, J.; Yin, K. Y. The Influence of Students' Concept of Mole, Problem Representation Ability and Mathematical Ability on Stoichiometry Problem Solving. *Scott. J. Arts Soc. Sci. Sci. Stud.* **2014**, 3.

(38) Lythcott, J. Problem Solving and Requisite Knowledge of Chemistry. *J. Chem. Educ.* **1990**, *67* (3), 248.

(39) Case, J. M.; Fraser, D. M. An Investigation into Chemical Engineering Students' Understanding of the Mole and the Use of Concrete Activities to Promote Conceptual Change. *Int. J. Sci. Educ.* **1999**, *21* (12), 1237–1249.

(40) Nurrenbern, S. C.; Pickering, M. Concept Learning versus Problem Solving: Is There a Difference? *J. Chem. Educ.* **1987**, *64* (6), 508.

(41) Nakhleh, M. B.; Mitchell, R. C. Concept Learning versus Problem Solving: There Is a Difference. *J. Chem. Educ.* **1993**, *70* (3), 190.

(42) Nakhleh, M. B. Are Our Students Conceptual Thinkers or Algorithmic Problem Solvers? Identifying Conceptual Students in General Chemistry. *J. Chem. Educ.* **1993**, *70* (1), 52.

(43) Sawrey, B. A. Concept Learning versus Problem Solving: Revisited. *J. Chem. Educ.* **1990**, *67* (3), 253.

(44) Freeman, S.; Eddy, S. L.; McDonough, M.; Smith, M. K.; Okoroafor, N.; Jordt, H.; Wenderoth, M. P. Active Learning Increases Student Performance in Science, Engineering, and Mathematics. *Proc. Natl. Acad. Sci.* **2014**, *111* (23), 8410–8415.

(45) Labov, J. B.; Singer, S. R.; George, M. D.; Schweingruber, H. A.; Hilton, M. L. Effective Practices in Undergraduate STEM Education Part 1: Examining the Evidence. *CBE-Life Sci. Educ.* **2009**, *8* (3), 157–161.

(46) Bunce, D. M. Introduction to Symposium on "Lecture and Learning: Are They Compatible?" *J. Chem. Educ.* **1993**, *70* (3), 179.

(47) Miller, T. L. Demonstration-Exploration-Discussion: Teaching Chemistry with Discovery and Creativity. *J. Chem. Educ.* **1993**, *70* (3), 187.

(48) Bodner, G. M. Why Changing the Curriculum May Not Be Enough. *J. Chem. Educ.* **1992**, *69* (3), 186.

(49) Bodner, G. M.; Metz, P. A.; Tobin, K. Cooperative Learning: An Alternative to Teaching at a Medieval University. *Aust. Sci. Teach. J.* **1997**, *43* (1), 23–28.

(50) Zoller, U. Are Lecture and Learning Compatible? Maybe for LOCS: Unlikely for HOCS. *J. Chem. Educ.* **1993**, *70* (3), 195.

(51) Hanson, D. M.; Wolfskill, T. Improving the Teaching/Learning Process in General Chemistry: Report on the 1997 Stony Brook General Chemistry Teaching Workshop. *J. Chem. Educ.* **1998**, *75* (2), 143.

(52) Spencer, J. N. New Directions in Teaching Chemistry: A Philosophical and Pedagogical Basis. *J. Chem. Educ.* **1999**, *76* (4), 566–null.

(53) Farrell, J. J.; Moog, R. S.; Spencer, J. N. A Guided-Inquiry General Chemistry Course. *J. Chem. Educ.* **1999**, *76* (4), 570.

(54) Spencer, J. N. New Approaches to Chemistry Teaching. 2005 George C. Pimentel Award. *J. Chem. Educ.* **2006**, *83* (4), 528–null.

(55) Weaver, G. C.; Sturtevant, H. G. Design, Implementation, and Evaluation of a Flipped Format General Chemistry Course. *J. Chem. Educ.* **2015**, *92* (9), 1437–1448.

(56) Muzyka, J. L. ConfChem Conference on Flipped Classroom: Just-in-Time Teaching in Chemistry Courses with Moodle. *J. Chem. Educ.* **2015**, *92* (9), 1580–1581.

(57) Butzler, K. B. ConfChem Conference on Flipped Classroom: Flipping at an Open-Enrollment College. *J. Chem. Educ.* **2015**, *92* (9), 1574–1576.

(58) Rossi, R. D. ConfChem Conference on Flipped Classroom: Improving Student Engagement in Organic Chemistry Using the Inverted Classroom Model. *J. Chem. Educ.* **2015**, *92* (9), 1577–1579.

(59) Haile, J. D. ConfChem Conference on Flipped Classroom: Using a Blog To Flip a Classroom. *J. Chem. Educ.* **2015**, *92* (9), 1572–1573.

(60) Trogden, B. G. ConfChem Conference on Flipped Classroom: Reclaiming Face Time—How an Organic Chemistry Flipped Classroom Provided Access to Increased Guided Engagement. *J. Chem. Educ.* **2015**, *92* (9), 1570–1571.

(61) Hartman, J. R.; Dahm, D. J.; Nelson, E. A. ConfChem Conference on Flipped Classroom: Time-Saving Resources Aligned with Cognitive Science To Help Instructors. *J. Chem. Educ.* **2015**, *92* (9), 1568–1569.

(62) Seery, M. K. ConfChem Conference on Flipped Classroom: Student Engagement with Flipped Chemistry Lectures. *J. Chem. Educ.* **2015**, *92* (9), 1566–1567.

(63) Luker, C.; Muzyka, J.; Belford, R. Introduction to the Spring 2014 ConfChem on the Flipped Classroom. *J. Chem. Educ.* **2015**, *92* (9), 1564–1565.

(64) Lage, M. J.; Platt, G. J.; Treglia, M. Inverting the Classroom: A Gateway to Creating an Inclusive Learning Environment. *J. Econ. Educ.* **2000**, *31* (1), 30–43.

(65) Baker, J. W. The "classroom Flip": Using Web Course Management Tools to Become the Guide by the Side. Selected Papers from the 11th International Conference on College Teaching and Learning. **2000**. 9-17

(66) Smith, J. D. Student Attitudes toward Flipping the General Chemistry Classroom. *Chem. Educ. Res. Pract.* **2013**, *14* (4), 607–614.

(67) Hibbard, L.; Sung, S.; Wells, B. Examining the Effectiveness of a Semi-Self-Paced Flipped Learning Format in a College General Chemistry Sequence. *J. Chem. Educ.* **2016**, *93* (1), 24–30.

(68) Ryan, M. D.; Reid, S. A. Impact of the Flipped Classroom on Student Performance and Retention: A Parallel Controlled Study in General Chemistry. *J. Chem. Educ.* **2016**, *93* (1), 13–23.

(69) Hanson, D.; Wolfskill, T. Process Workshops - A New Model for Instruction. *J. Chem. Educ.* **2000**, *77* (1), 120.

(70) Black, K. A. What To Do When You Stop Lecturing: Become a Guide and a Resource. *J. Chem. Educ.* **1993**, *70* (2), 140.

(71) Wijtmans, M.; van Rens, L.; van Muijlwijk-Koezen, J. E. Activating Students' Interest and Participation in Lectures and Practical Courses Using Their Electronic Devices. *J. Chem. Educ.* **2014**, *91* (11), 1830–1837.

(72) Lee, A. W. M.; Ng, J. K. Y.; Wong, E. Y. W.; Tan, A.; Lau, A. K. Y.; Lai, S. F. Y. Lecture Rule No. 1: Cell Phones ON, Please! A Low-Cost Personal Response System for Learning and Teaching. *J. Chem. Educ.* **2013**, *90* (3), 388–389.

(73) King, D. B. Using Clickers To Identify the Muddiest Points in Large Chemistry Classes. *J. Chem. Educ.* **2011**, *88* (11), 1485–1488.

(74) Shaver, M. P. Using Low-Tech Interactions in the Chemistry Classroom To Engage Students in Active Learning. *J. Chem. Educ.* **2010**, *87* (12), 1320–1323.

(75) Pienta, N. J. A "Flipped Classroom" Reality Check. *J. Chem. Educ.* **2016**, *93* (1), 1–2.

(76) Finkel, Donald. L. *Teaching with your mouth shut.* **2000**. Portsmouth, NH: Boynton/Cook.

(77) Novick, S.; Menis, J. A Study of Student Perceptions of the Mole Concept. *J. Chem. Educ.* **1976**, *53* (11), 720.

(78) Driver, R.; Erickson, G. Theories-in-Action: Some Theoretical and Empirical Issues in the Study of Students' Conceptual Frameworks in Science. **1983**.

(79) Garnett, P. J.; Garnett, P. J.; Hackling, M. W. Students' Alternative Conceptions in Chemistry: A Review of Research and Implications for Teaching and Learning. **1995**.

(80) Gabel, D. L. Use of the Particle Nature of Matter in Developing Conceptual Understanding. *J. Chem. Educ.* **1993**, *70* (3), 193.

(81) Bunce, D. M. Teaching Is More Than Lecturing and Learning Is More Than Memorizing. 2007 James Flack Norris Award. *J. Chem. Educ.* **2009**, *86* (6), 674.